MACHINE COMPONENT ANALYSIS WITH MATLAB®

DAN B. MARGHITU

MIHAI DUPAC

Butterworth-Heinemann
An imprint of Elsevier

Butterworth-Heinemann is an imprint of Elsevier
The Boulevard, Langford Lane, Kidlington, Oxford OX5 1GB, United Kingdom
50 Hampshire Street, 5th Floor, Cambridge, MA 02139, United States

Notices

Knowledge and best practice in this field are constantly changing. As new research and experience
broaden our understanding, changes in research methods, professional practices, or medical
treatment may become necessary.

Practitioners and researchers must always rely on their own experience and knowledge in evaluating
and using any information, methods, compounds, or experiments described herein. In using such
information or methods they should be mindful of their own safety and the safety of others,
including parties for whom they have a professional responsibility.

To the fullest extent of the law, neither the Publisher nor the authors, contributors, or editors,
assume any liability for any injury and/or damage to persons or property as a matter of products
liability, negligence or otherwise, or from any use or operation of any methods, products,
instructions, or ideas contained in the material herein.

Library of Congress Cataloging-in-Publication Data
A catalog record for this book is available from the Library of Congress

British Library Cataloguing-in-Publication Data
A catalogue record for this book is available from the British Library

ISBN: 978-0-12-804229-8

For information on all Butterworth-Heinemann publications
visit our website at https://www.elsevier.com/books-and-journals

Working together
to grow libraries in
developing countries

www.elsevier.com • www.bookaid.org

Publisher: Matthew Deans
Acquisition Editor: Brian Guerin
Editorial Project Manager: Isabella C. Silva
Production Project Manager: Nirmala Arumugam
Designer: Mark Rogers

Typeset by VTeX

Contents

Preface

This book is intended as a supplement for courses in machine component design and as a reference for mechanical engineers. The book uses MATLAB® as a tool to analyze and solve machine component problems.

The solutions of the problems are obtained analytically and numerically using MATLAB. Many figures are generated with MATLAB programs. Specific functions dealing with machine components are created. The book will assist the undergraduate and advanced undergraduate students interested in machine element analysis. The project can be used for classroom instruction and it can be used for a self-study and can also be offered as distance learning.

The chapters of the book are: stress and deflection, fatigue failure, screws, rolling bearings, lubrication and sliding bearings, and spur gears.

Dan B. Marghitu and Mihai Dupac

CHAPTER ONE

Stress and deflection

1.1. Stress components

In the design process, the uniform distribution of stresses is usually considered, that is, the results of forces and moments applied to an element represent pure shear or pure tension. If a force F (tension) acts at the ends of a straight bar its line of action incorporates the centroid of the section. If one piece is cut and removed from the bar (made of a homogeneous material), one can replace its effect through a force $F = \sigma A$ uniformly distributed at the cut. The normal stress σ can then be expressed as

$$\sigma = \frac{F}{A}, \tag{1.1}$$

where A is the bar cross-sectional area. For an element in shear the uniform shear stress distribution is

$$\tau = \frac{F}{A}. \tag{1.2}$$

Stress elements represent a convenient way to show stresses acting at some point of a body. A three-dimensional Cartesian stress element with the normal stresses σ_x, σ_y, and σ_z, and the shear stresses τ_{xz}, τ_{zx}, τ_{xy}, τ_{yx}, τ_{yz}, and τ_{zy}, is shown in Fig. 1.1A. For the shear stresses, using the static equilibrium, results in

$$\tau_{xz} = \tau_{zx}, \quad \tau_{yx} = \tau_{xy}, \quad \tau_{zy} = \tau_{yz}. \tag{1.3}$$

If the stresses σ_x, σ_y, and σ_z – called either tensile stresses or tension – are oriented as shown in Fig. 1.1A, the stresses are considered positive. The subscripts used in the definition of the normal stresses represent the normal direction to the surface.

The shear stresses acting in the positive direction of the reference axis are considered positive. The first and second subscripts of the shear stress denote the axes to which it is perpendicular (and subsequently the face on which the stress acts) and respectively parallel.

For the stress element shown in Fig. 1.1B, where only the x and y faces are subject to stresses, $\sigma_z = 0$ and $\tau_{yz} = \tau_{zy} = \tau_{xz} = \tau_{zx} = 0$, and σ_x and σ_y

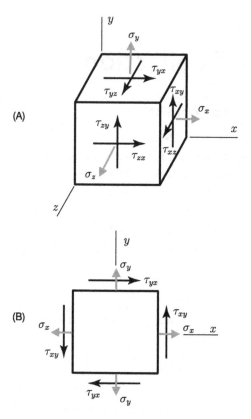

Figure 1.1 (A) Three-dimensional stress element and (B) planar element. From Budynas–Nisbett: Shigley's Mechanical Engineering Design, Eighth Edition, McGraw-Hill, 2006. Used with permission from McGraw Hill Inc.

act in the positive direction. The shear stresses acting in the clockwise (cw) direction are considered positive, otherwise negative and acting counter-clockwise (ccw).

Many times it is desirable to calculate stresses on an inclined (or rotated) section acting at an angle ϕ (Fig. 1.2). The stresses τ and σ acting on an inclined plane (section) can be computed considering the equilibrium equations for the force components caused by the stresses by using the formulas

$$\sigma = \frac{\sigma_x + \sigma_y}{2} + \frac{\sigma_x - \sigma_y}{2}\cos 2\phi + \tau_{xy}\sin 2\phi, \qquad (1.4)$$

$$\tau = \frac{\sigma_y - \sigma_x}{2}\sin 2\phi + \tau_{xy}\cos 2\phi. \qquad (1.5)$$

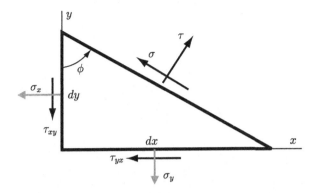

Figure 1.2 Stresses σ and τ on an oblique plane. From Budynas–Nisbett: Shigley's Mechanical Engineering Design, Eighth Edition, McGraw-Hill, 2006. Used with permission from McGraw Hill Inc.

Considering that the derivative with respect to the angle ϕ of Eq. (1.4) equals zero, one can write

$$\frac{d\tau}{d\phi} = \frac{\sigma_x - \sigma_y}{2}(-2\sin 2\phi) + \tau_{xy}(2\cos 2\phi) = 0, \tag{1.6}$$

or equivalently,

$$\tan 2\phi = \frac{2\tau_{xy}}{\sigma_x - \sigma_y}. \tag{1.7}$$

The two solutions of Eq. (1.7) as functions of ϕ represent the *principal stresses* σ_1 and σ_2, respectively named minimum and maximum stresses. The angles ϕ are called *principal angles*, and the related (or matching) directions that are perpendicular to each other are called *principal directions*.

Setting the derivative of Eq. (1.5) to zero, one obtains

$$\tan 2\phi = -\frac{\sigma_x - \sigma_y}{2\tau_{xy}}. \tag{1.8}$$

Solving Eq. (1.8), one can obtain the angles 2ϕ which represent the extreme values of the shear stress τ. Rewriting Eq. (1.7) as

$$(\sigma_x - \sigma_y)\sin 2\phi = 2\tau_{xy}\cos 2\phi,$$

one obtains

$$\sin 2\phi = \frac{2\tau_{xy}\cos 2\phi}{\sigma_x - \sigma_y}. \tag{1.9}$$

Combining Eqs. (1.5) and (1.9), one can write

$$\tau = -\frac{\sigma_x - \sigma_y}{2} \frac{2\tau_{xy}\cos 2\phi}{\sigma_x - \sigma_y} + \tau_{xy}\cos 2\phi = 0, \tag{1.10}$$

that is, the shear stress in the principal directions is negligible.

Combining Eqs. (1.4) and (1.8), one gets

$$\sigma = \frac{\sigma_x + \sigma_y}{2}, \tag{1.11}$$

that is, the normal stresses related with the maximum shear stresses are equal.

Using Eq. (1.7), one can calculate

$$\sin 2\phi = \frac{\tau_{xy}}{\sqrt{\tau_{xy}^2 + \left(\frac{\sigma_x - \sigma_y}{2}\right)^2}}, \quad \cos 2\phi = \frac{\sigma_x - \sigma_y}{2\sqrt{\tau_{xy}^2 + \left(\frac{\sigma_x - \sigma_y}{2}\right)^2}}. \tag{1.12}$$

From Eqs. (1.4) and (1.12) one can obtain the two principal stresses σ_1 and σ_2 as

$$\sigma_1 = \frac{\sigma_x + \sigma_y}{2} + \sqrt{\tau_{xy}^2 + \left(\frac{\sigma_x - \sigma_y}{2}\right)^2},$$

$$\sigma_2 = \sigma_x + \sigma_y - \sigma_1 = \frac{\sigma_x + \sigma_y}{2} - \sqrt{\tau_{xy}^2 + \left(\frac{\sigma_x - \sigma_y}{2}\right)^2}, \tag{1.13}$$

or equivalently,

$$\sigma_{1,2} = \frac{\sigma_x + \sigma_y}{2} \pm \sqrt{\tau_{xy}^2 + \left(\frac{\sigma_x - \sigma_y}{2}\right)^2}, \tag{1.14}$$

where σ_2 and σ_1 are respectively the smaller and larger principal stresses. The planes with no shear stress (when shear stresses are zero) are called *principal planes*. Similarly, the shear stresses can be written as

$$\tau_1, \tau_2 = \pm\sqrt{\tau_{xy}^2 + \left(\frac{\sigma_x - \sigma_y}{2}\right)^2}. \tag{1.15}$$

The state of stress can be graphically represented using the *Mohr's circle diagram* method (named after German civil engineer Otto Mohr) shown

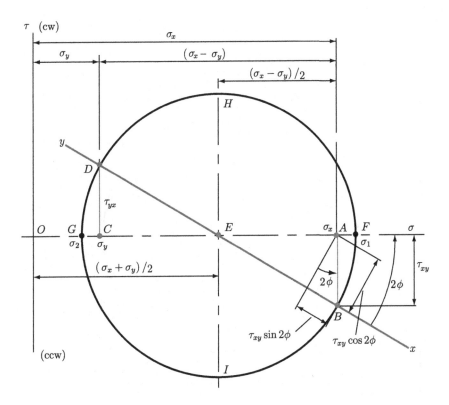

Figure 1.3 Mohr's circle diagram.

in Fig. 1.3. The idea behind this graphical representation is that all the possible values σ and τ in Eqs. (1.4) and (1.5) for a given state of stress can be obtained by varying the angle θ from 0 to 360°. The normal and the shear stresses are plotted on the abscissa and (respectively) ordinate axis, while tensile (σ_x and σ_y shown in Fig. 1.3) and normal compressive stresses are considered positive and (respectively) negative. The shear stresses are considered positive or negative if their orientation is clockwise (cw) or counterclockwise (ccw).

For the Mohr's circle diagram presented in Fig. 1.3, the tensile stresses σ_y and σ_x are represented by OA and OC, respectively, τ_{xy} by AB, and τ_{yx} by CD. The Mohr's circle center is located at E, points B and D having coordinates (σ_x, τ_{xy}) and (σ_y, τ_{yx}) represent the stress conditions on the x- and y-face, respectively. Points B and D are the opposite ends of the Mohr's circle diameter, therefore $2\phi = 180°$, and the angle (on the stress element) between x and y is $\phi = 90°$. The principal normal stresses σ_1 (maximum)

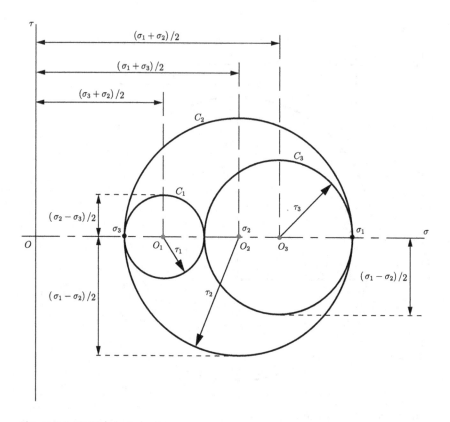

Figure 1.4 3D Mohr's circle diagram.

and σ_2 (minimum) are located at points F and G, while the shear stresses' extreme values are located at H and I.

Considering a three-dimensional stress element, an orientation having all the shear stress elements/components equal to zero and the principal directions normal to the faces can be considered. For a stress element, the three principal directions are related to the stresses σ_1, σ_2, and σ_3. The components needed to specify a given state of stress of a three-dimensional element are τ_{xy}, τ_{yz}, τ_{zx}, and σ_x, σ_y, σ_z.

To devise Mohr's circle for a three-dimensional state of stresses (Fig. 1.4), the values σ_1, σ_2, and σ_3 representing the principal stresses are ordered, that is, $\sigma_1 > \sigma_2 > \sigma_3$. The *principal shear stresses* (Fig. 1.4) τ_1, τ_2, and τ_3 are computed using

$$\tau_1 = \frac{1}{2}(\sigma_2 - \sigma_3), \quad \tau_2 = \frac{1}{2}(\sigma_1 - \sigma_3), \quad \tau_3 = \frac{1}{2}(\sigma_1 - \sigma_2). \quad (1.16)$$

The Mohr's circles shown in Fig. 1.4 represent the normal and shearing stresses for rotation about the principal axes. If $\tau_{max} = \tau_2$, then $\sigma_1 > \sigma_2 > \sigma_3$.

Elastic strain

The amount of elongation obtained by applying tensile load to a straight bar is named *total strain*. Also known as unit deformation, the bar elongation per unit length can be expressed as

$$\epsilon = \frac{\delta}{l}, \tag{1.17}$$

where l is the bar length, ϵ is the Greek symbol used to designate *strain*, δ is the total strain (or bar elongation), and ϵ is dimensionless.

The angular change in the right angle of an rectangular element (under the action of pure shear stresses) named *shear strain* can be written as

$$\gamma = \frac{\delta_s}{l}. \tag{1.18}$$

Elasticity is the ability of a material to regain its original shape/geometry once the applied load is withdrawn. Considering Hooke's law of elasticity, that is, for small deformations, the stress is directly proportional to the strain (that produced it), and one can write

$$\sigma = E\epsilon, \qquad \tau = G\gamma, \tag{1.19}$$

where the Young's modulus E measures the material resistance to being elastically deformed, and the *shear modulus* G – named also *modulus of rigidity* – measures the material resistance to being elastically deformed.

For axial tension or compression, the total deformation δ of a bar can be calculated from Eqs. (1.1), (1.17), and (1.19) as

$$\delta = \frac{Fl}{AE}. \tag{1.20}$$

When a tensile load is applied to a bar (Fig. 1.5), there is an increase in length but also a decrease in width, that is, axial and lateral strains – proportional to each other – occur. The negative lateral to axial strain ratio – named *Poisson's ratio* – is given by

$$\nu = -\frac{\text{lateral strain}}{\text{axial strain}}. \tag{1.21}$$

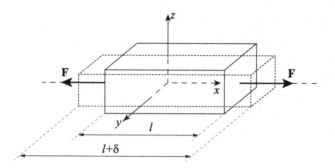

Figure 1.5 Tensile load applied to a bar and its deformation.

If the material is subject to stress only in the x direction, the corresponding strain is $\epsilon_x = \dfrac{\sigma_x}{E}$. If the stress σ_y is acting in the y direction, the induced strain is $\epsilon_x = -\nu\epsilon_y = -\nu\dfrac{\sigma_y}{E}$, where ν is the Poisson ratio expressed as

$$\nu = -\frac{\epsilon_y}{\epsilon_x} = -\frac{\epsilon_z}{\epsilon_x}, \tag{1.22}$$

ϵ_x is the axial strain, ϵ_y and ϵ_z are the lateral strains, i.e., strains in the normal directions. If the bar is subjected to both stresses at once ($\sigma_z = 0$), the corresponding strains can be expressed by

$$\begin{aligned}
\epsilon_x &= \frac{1}{E}\left(\sigma_x - \nu\sigma_y\right), \\
\epsilon_y &= \frac{1}{E}\left(\sigma_y - \nu\sigma_x\right), \\
\epsilon_z &= -\frac{1}{E}\left(\nu\sigma_x - \nu\sigma_y\right).
\end{aligned} \tag{1.23}$$

For the 3D case when the bar is subjected to the stresses σ_x, σ_y, and σ_z at once, the corresponding strains can be expressed by

$$\begin{aligned}
\epsilon_x &= \frac{1}{E}\left(\sigma_x - \nu\sigma_y - \nu\sigma_z\right), \\
\epsilon_y &= \frac{1}{E}\left(\sigma_y - \nu\sigma_x - \nu\sigma_z\right), \\
\epsilon_z &= \frac{1}{E}\left(\sigma_z - \nu\sigma_x - \nu\sigma_y\right).
\end{aligned} \tag{1.24}$$

The modulus of elasticity E, Poisson's ratio v and shear modulus G are related by

$$G = \frac{E}{2(1 + v)}. \tag{1.25}$$

The bulk modulus of elasticity given by

$$K = \frac{E}{3(1 - 2v)} \tag{1.26}$$

represents the resistance of a material to compression under uniform loading.

Shear and moment

The simply supported beam in Fig. 1.6A is loaded by a transversal force F acting downwards. The reactions at supports, denoted by R_1 and R_2, are directed on the positive direction of the y-axis and are considered positive. The beam is sectioned at a distance x from its left end, and its left-hand part is considered free. To secure equilibrium, a shear force denoted by V and a bending moment denoted by M should be applied at the cutting (Fig. 1.6B).

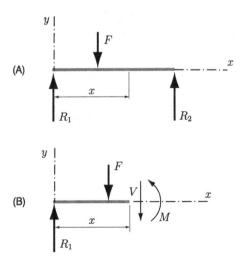

Figure 1.6 (A) Simply supported beam under the action of concentrated forces, and (B) free-body diagram of the sectioned beam. From Dan B. Marghitu, Kinematic Chains and Machine elements Design, Elsevier, 2005.

When bending is due to a uniformly distributed load w, the equilibrium of the forces along the y direction (for a differential beam element) gives

$$\frac{dV}{dx} = -w, \tag{1.27}$$

where w is expressed in units of force per unit length. By equilibrium of moments, one can relate the bending moment and the shear force by

$$V = \frac{dM}{dx}. \tag{1.28}$$

For a differential element of the beam, one can integrate Eq. (1.27) to obtain V as

$$\int dV = \int -w\,dx, \tag{1.29}$$

and then integrate Eq. (1.28) to obtain M as

$$\int dM = \int V\,dx, \tag{1.30}$$

where the integration constants can be obtained from the values of the bending moment and shear force at the ends of the beam. The above equations confirm that the changes in shear force and bending moment for a differential element represent the area of the loading and shear force diagrams of the element, respectively.

Normal stresses in pure bending

To calculate the normal stresses under pure bending, the following assumptions about the beam are taken into consideration:

1. The beam is made of an isotropic and homogeneous material which obeys Hooke's law.

2. The beam is straight, with its axis of symmetry in the plane of bending.

3. The beam has a constant cross-section.

A graphical representation of a beam section – under the action of a bending moment $\mathbf{M}_z$ – is shown in Fig. 1.7. The beam *neutral plane* (represented by the xz plane) is coincident with the zero strain elements of the beam, and its *neutral* and *symmetry axes* are coincident with x and y axes, respectively.

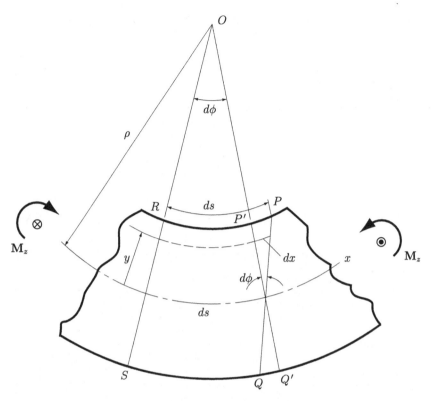

Figure 1.7 Normal stress in flexure. From J. Shigley, C.R. Mischke, Mechanical Engineering Design, 5th edition, McGraw Hill Press, New York, 1989. Used with permission from McGraw Hill Inc.

If a (positive) moment is applied to the beam (Fig. 1.7), the beam neutral axis and upper surface will bend downward, that is, the elements at the top will be in compression while the elements below the neutral axis will be in extension. Initially parallel to the section RS, section PQ rotates by the angle $d\phi$ (Fig. 1.7) to its final orientation $P'Q'$. Considering the geometry of the beam (Fig. 1.7) with ds the neutral axis differential element length, $d\phi$ the adjacent angle, and ρ the radius of curvature, one can write

$$\frac{1}{\rho} = \frac{d\phi}{ds}. \tag{1.31}$$

The strain due to the beam deformation can be expressed by

$$\epsilon = -\frac{dx}{ds} = -\frac{y \, d\phi}{ds} = -\frac{y}{\rho}, \tag{1.32}$$

where $dx = y\,d\phi$. From Eqs. (1.19) and (1.32), the associated stress can be calculated as

$$\sigma = E\epsilon = -\frac{Ey}{\rho}. \tag{1.33}$$

Considering an element of area dA, one can express the force acting on the element by

$$dF = \sigma\,dA. \tag{1.34}$$

By integration one can obtain

$$\int \sigma\,dA = -\frac{E}{\rho}\int y\,dA = 0. \tag{1.35}$$

At equilibrium, the internal and external moments σ and $\mathbf{M}_z = M\mathbf{k}$ should be the same, that is,

$$M = \int y\sigma\,dA = \frac{E}{\rho}\int y^2\,dA = \frac{EI}{\rho}, \tag{1.36}$$

or equivalently,

$$\frac{1}{\rho} = \frac{M}{EI}, \tag{1.37}$$

where $I = \int y^2\,dA$ is the second moment of area (with respect to the z axis). Eliminating ρ from Eqs. (1.33) and (1.37) yields

$$\sigma = -\frac{My}{I}, \tag{1.38}$$

that is, the stress is proportional to the product My as shown in Fig. 1.8.

The maximum stress can now be expressed as

$$\sigma_{max} = \frac{Mc}{I} = \frac{M}{I/c} = \frac{M}{Z_{sm}}, \tag{1.39}$$

where $c = y_{max}$ and $Z_{sm} = I/c$ is the *section modulus*.

For *asymmetrical sections*, the previous results may be considered if and only if the section beam principal axis coincides with the plane of bending. Using Eqs. (1.34) and (1.35), the moment of the force (for a differential element dA) about the y-axis can be expressed by

$$M_y = \int z\,dF = \int z\sigma\,dA = \int z\frac{-Ey}{\rho}\,dA = -\frac{E}{\rho}\int yz\,dA, \tag{1.40}$$

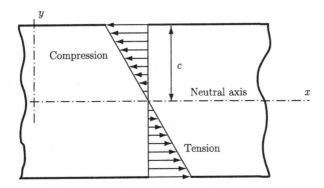

Figure 1.8 Bending stress in flexure. From Budynas–Nisbett: Shigley's Mechanical Engineering Design, Eighth Edition, McGraw-Hill, 2006. Used with permission from McGraw Hill Inc.

where the integral is over the section. If the bending moment acts in the plane of the principal axis, one can deduce that $I_{yz} = \int yz \, dA = 0$ and $M_y = 0$. Hence, the developed relations can be applied to asymmetrical sections if and only if $I_{yz} = 0$.

Shear stresses in beams

The beam of constant cross-section in Fig. 1.9 is subject to a bending moment $\mathbf{M}_z = M\mathbf{k}$ and a shear force $\mathbf{V} = V\mathbf{J}$, where the unit vectors $\mathbf{J}$ and $\mathbf{k}$ relate to the y- and z-axes. The bending moment $\mathbf{M}_z = M\mathbf{k}$ and a shear force $\mathbf{V} = V\mathbf{J}$ are related by Eq. (1.28). Consider next the beam element of length dx in Fig. 1.9. The bending moment is not constant along the x-axis due to the shear force, that is, the bending moments M and $M + dM$ act at the beginning and end side of the beam element. The bending moments that acts at the sections of the beam element produce the normal stresses σ and $\sigma + d\sigma$, respectively. The force created by the normal stress σ can be expressed by

$$\mathbf{F}_b = F_b \mathbf{1}, \quad \text{where} \quad F_b = \int_{y_1}^{c} \sigma \, dA, \tag{1.41}$$

and consequently, the force created by $\sigma + d\sigma$ is given by

$$\mathbf{F}_e = -F_e \mathbf{1}, \quad \text{where} \quad F_e = \int_{y=y_1}^{y=c} (\sigma + d\sigma) \, dA, \tag{1.42}$$

where dA is the area of a small element of the face.

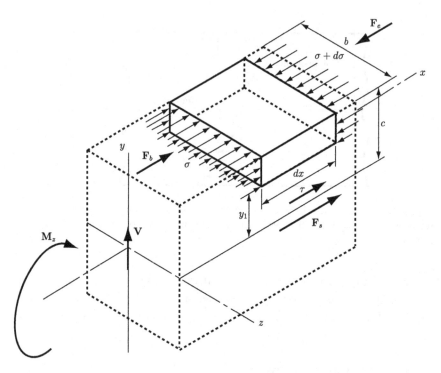

Figure 1.9 Shear stresses. From Dan B. Marghitu, Kinematic Chains and Machine elements Design, Elsevier, 2005.

Using $\sigma = \dfrac{My}{I}$, the forces acting on the beginning side and on the end face can be calculated as

$$F_b = \frac{M}{I} \int_{y_1}^{c} y\, dA, \tag{1.43}$$

$$F_e = \frac{M + dM}{I} \int_{y_1}^{c} y\, dA. \tag{1.44}$$

Since $F_e > F_b$, the resultant $R_F = F_e - F_b$ tends to slide the element in the $-x$ direction. To secure equilibrium, the shear force must act in the opposite direction to the resultant. The shear force can be expressed by

$$\mathbf{F}_s = F_s \mathbf{1}, \quad \text{where} \quad F_s = b\tau\, dx, \tag{1.45}$$

where τ is the shear stress and b is the element width.

At equilibrium, the sum of forces in the x direction equals zero, that is,

$$\sum F_x = F_s + F_b - F_e = 0$$
$$\Leftrightarrow \quad F_s = F_e - F_b$$
$$\Leftrightarrow \quad F_s = \frac{M + dM}{I} \int_{y_1}^{c} y \, dA - \frac{M}{I} \int_{y_1}^{c} y \, dA$$
$$\Leftrightarrow \quad F_s = \frac{dM}{I} \int_{y_1}^{c} y \, dA. \tag{1.46}$$

Substituting Eqs. (1.28) and (1.45) into Eq. (1.46), one can calculate the shear stress as

$$\tau = \frac{1}{Ib} \frac{dM}{dx} \int_{y_1}^{c} y \, dA = \frac{V}{Ib} \int_{y_1}^{c} y \, dA, \tag{1.47}$$

or in a much simpler form using the first moment of area Q (about the neutral axis) as

$$\tau = \frac{VQ}{Ib}. \tag{1.48}$$

For the beam of rectangular cross-section in Fig. 1.10, the first moment of area Q becomes

$$Q = \int_{y_1}^{c} y \, dA = b \int_{y_1}^{c} y \, dy = \frac{1}{2} b y^2 \bigg|_{y_1}^{c} = \frac{1}{2} b (c^2 - y_1^2), \tag{1.49}$$

where $dA = b \, dy$, the element dA is at a distance y above the neutral axis, $\mathbf{V} = V\mathbf{J}$ is the shear force, $\mathbf{M}_z = M\mathbf{k}$ is the bending moment and σ is the normal stress at the beam cross-section. Using Eq. (1.49), the shear stress in Eq. (1.48) can be expressed by

$$\tau = \frac{V}{2I}(c^2 - y_1^2). \tag{1.50}$$

The second moment of area in the case of a rectangular section can be expressed as

$$I = \frac{bh^2}{12} = \frac{Ac^2}{3}, \tag{1.51}$$

where $A = bh = 2bc$. Eqs. (1.50) and (1.51) give

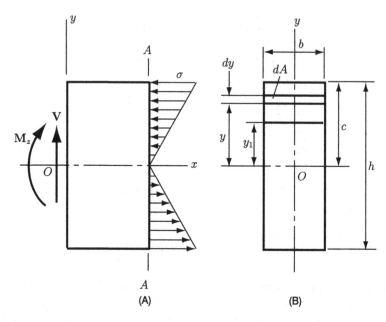

Figure 1.10 Stresses in a beam with rectangular cross-section: (A) side view, and (B) cross-section view. From Budynas–Nisbett: Shigley's Mechanical Engineering Design, Eighth Edition, McGraw-Hill, 2006. Used with permission from McGraw Hill Inc.

$$\tau = \frac{3V}{2A}\left(1 - \frac{y_1^2}{c^2}\right) = C\frac{V}{A},$$ (1.52)

where the value of C is [4]

$$C = \begin{cases} 1.50 & \text{for } y_1 = 0, \\ 1.44 & \text{for } y_1 = 0.2\,c, \\ 1.26 & \text{for } y_1 = 0.4\,c, \\ 0.96 & \text{for } y_1 = 0.6\,c, \\ 0 & \text{for } y_1 = c. \end{cases}$$ (1.53)

A maximum shear stress $\tau_{max} = \dfrac{3V}{2A}$ can be obtained at the neutral axis. The maximum flexural shear stress for rectangular, circular, hollow round and structural shapes can be expressed as in [27] by

$$\tau_{rectangular} = \frac{3V}{2A}, \qquad \tau_{circular} = \frac{4V}{3A},$$
$$\tau_{hollow} = \frac{2V}{A}, \qquad \tau_{structural} = \frac{V}{A_{web}}.$$ (1.54)

1.2. Deflection

When an element does not change its geometry under the action of an external force, it is called *rigid*. Otherwise it is called a *flexible* element which deflects, bends or twists under the action of an external force, torque or moment. If a material regains its initial shape after having been deformed, it is called *elastic*, and this property is called *elasticity*.

Consider next the deflection of a simply supported beam (Fig. 1.11) under a loading transversal force F. If the elastic limit of the material – due to the deflection – is not exceeded, the beam can be represented as a *linear spring*.

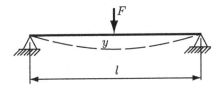

Figure 1.11 Beam described as a linear spring. From Budynas–Nisbett: Shigley's Mechanical Engineering Design, Eighth Edition, McGraw-Hill, 2006. Used with permission from McGraw Hill Inc.

The spring rate of a loaded bar can be expressed using Eq. (1.20) and the *spring constant* definition by

$$k = \frac{F}{y} = \frac{AE}{l}, \qquad (1.55)$$

where l is the bar length, E is the Young's modulus, F is the applied force, and A is the area of the bar cross-section.

The curvature of a planar beam can be expressed by

$$\frac{1}{\rho} = \frac{\dfrac{d^2 y}{dx^2}}{\sqrt{\left[1 + \left(\dfrac{dy}{dx}\right)^2\right]^3}}, \qquad (1.56)$$

where $y(x)$ is the beam deflection. If the curve slope $\theta = \dfrac{dy}{dx}$ is very small, its square $\left(\dfrac{dy}{dx}\right)^2$ is negligible, and the denominator can be approximated by

$$\sqrt{\left[1+\left(\frac{dy}{dx}\right)^2\right]^3} \approx 1.$$

Consequently, the curvature can be approximated by

$$\frac{1}{\rho} = \frac{d^2y}{dx^2}. \tag{1.57}$$

The beam curvature can be related to the beam external moment M by

$$\frac{1}{\rho} = \frac{M}{EI}, \tag{1.58}$$

where E is the modulus of elasticity and I the second moment of area. From Eqs. (1.57) and (1.58), one can obtain

$$\frac{M}{EI} = \frac{d^2y}{dx^2}. \tag{1.59}$$

Differentiating Eq. (1.59) twice, one can write

$$\frac{V}{EI} = \frac{d^3y}{dx^3}, \tag{1.60}$$

$$\frac{q}{EI} = \frac{d^4y}{dx^4}, \tag{1.61}$$

where $V = \dfrac{dM}{dx}$ is the shear force.

Strain energy

Strain energy or *potential energy* defined as work done by an external force, torque or moment on a elastic member to distort its shape can be expressed as

$$U = \frac{1}{2}F\gamma = \frac{1}{2}\frac{F^2}{k}, \tag{1.62}$$

where $\gamma = \dfrac{F}{k}$ is the member deformation.

For tension and compression, the potential energy can be determined by

$$U = \frac{1}{2}F\delta = \frac{F^2l}{2AE}, \tag{1.63}$$

where the deformation δ is calculated using Eq. (1.20).

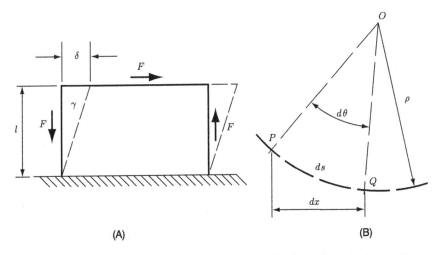

Figure 1.12 Strain energy: (A) direct shear and (B) bending. From Budynas–Nisbett: Shigley's Mechanical Engineering Design, Eighth Edition, McGraw-Hill, 2006. Used with permission from McGraw Hill Inc.

Fig. 1.12A shows a fixed element (by its side) in pure shear. The stain energy due to shear can be represented by

$$U = \frac{1}{2}F\delta = \frac{F^2 l}{2AG}, \tag{1.64}$$

where the deformation δ is calculated from the shear strain $\gamma = \dfrac{F}{AG}$ due to the applied force F.

The potential energy due to bending (Fig. 1.12B) can be expressed as

$$dU = \frac{1}{2}Md\theta = \frac{Mds}{2\rho}, \tag{1.65}$$

where $\rho = \dfrac{ds}{d\theta}$ is the radius of curvature, and the length of the elastic curve PQ is $l_{PQ} = ds$. Using Eq. (1.58), one can eliminate the radius of curvature from Eq. (1.65) and obtain

$$dU = \frac{M^2 ds}{2EI}. \tag{1.66}$$

Integrating Eq. (1.66), one obtains the beam strain energy due to bending as

$$U = \frac{1}{2} \int \frac{M^2}{EI} \, dx, \tag{1.67}$$

where $ds \approx dx$ (the case of small deflections).

Using Eqs. (1.63)–(1.64), one can obtain the strain energy u as

$$u = \frac{1}{2} \frac{\sigma^2}{E} \quad \text{(tension/compression)},$$

$$u = \frac{1}{2} \frac{\tau_{max}^2}{2G} \quad \text{(torsion)},$$

$$u = \frac{1}{2} \frac{\tau^2}{G} \quad \text{(direct shear)}.$$

In the case of shear loading, the strain energy of a beam can be approximated using Eq. (1.64) as

$$U = \int \frac{CV^2 dx}{2AG}, \tag{1.68}$$

where V is the shear force, and C is a correction factor [27]

$$C = \begin{cases} 2 & \text{for round or tubular cross-section,} \\ 1.5 & \text{for rectangular cross-section,} \\ 1.33 & \text{for rectangular cross-section,} \\ 1 & \text{for structural or box cross-section.} \end{cases} \tag{1.69}$$

Castigliano's theorem may represent a good approach to deflection analysis, provided the system is in equilibrium.

Castigliano's theorem. *The displacement corresponding to any force in a system subjected to small elastic displacements is equal to the first partial derivative of the total strain energy in the system with respect to that force, in the direction of displacement.*

Castigliano's theorem can be written as

$$\delta_i = \frac{\partial U}{\partial F_i}, \tag{1.70}$$

where δ_i is the deflection, F_i is the force, and U is the potential/strain energy at a certain point. As an straightforward example, Castigliano's theorem can be applied for the case of axial deflection. Considering the expressions of the strain energy in Eq. (1.63), one can obtain

$$\delta = \frac{\partial}{\partial F}\left(\frac{F^2 l}{2AE}\right) = \frac{Fl}{AE}.\qquad(1.71)$$

1.3. Examples

Example 1.1. A rigid plate B is fastened to bar A and to bar C as shown in Fig. 1.13. Bar C is also fastened to bar D which has its end fastened to a rigid support while the end of bar A is free. Bars A, C, and D have lengths l_A, l_C, l_D and diameters $d_A = d$, $d_C = d$, and $d_D = 1.5\,d$, respectively. Load F_1 is applied to the rigid plate B (distributed uniformly around the circumference of the rigid plate B), and load F_2 is applied at the centroid of the end cross-section of bar A as shown in Fig. 1.13.

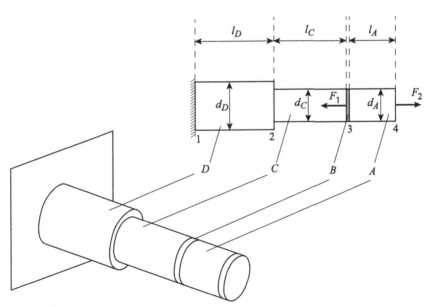

Figure 1.13 Axial bar under loading.

Determine the axial stresses in bars A, C, and D, deformations of the bars and total deformation of the system. For the numerical application use: $F_1 = 64\,000$ N, $F_2 = 192\,000$ N, $d = 45$ mm, $l_A = 170$ mm, $l_C = 144.5$ mm, $l_D = 255$ mm, and $E = 21\,(10^4)$ MPa.

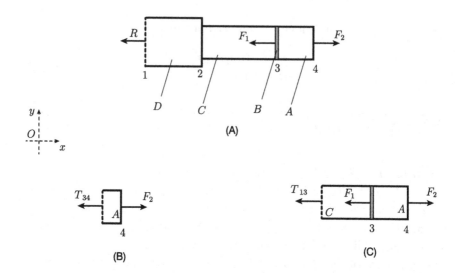

Figure 1.14 Free-body diagrams.

Solution

The bar is divided into components, and the internal forces are calculated passing sections through each component. Free-body diagrams are developed for portions of the bar.

The free-body diagram of the system, namely bar D, bar C, plate B, and bar A, contains only one unknown force, that is, reaction R at the rigid support, as shown in Fig. 1.14A. Reaction R, or the reaction force of the rigid support on bar D, is

$$\sum F_x^{A,B,C,D} = 0 \Leftrightarrow R + F_1 - F_2 = 0 \Leftrightarrow R = F_2 - F_1. \qquad (1.72)$$

The MATLAB® program starts with the declaration of variables:

```
clear all; clc; close all
syms F_1 F_2
syms d_A d_C d_D d E l_A l_C l_D
```

The internal forces are calculated for each portion of the bar. For the interval 3–4, the free-body diagram for bar A as shown in Fig. 1.14B gives

$$\sum F_x = 0 \Leftrightarrow T_{34} - F_2 = 0 \Leftrightarrow T_{34} = F_2. \qquad (1.73)$$

For the interval 1–3, the free-body diagram shown in Fig. 1.14C will give the equilibrium equation as

$$\sum F_x = 0 \Leftrightarrow T_{13} + F_1 - F_2 = 0 \Leftrightarrow T_{13} = F_2 - F_1, \qquad (1.74)$$

where T_{13} is the internal force for the interval 1–3.

The equilibrium equations are written in MATLAB with:

```
T_34 = F_2;
T_13 = F_2-F_1;
```

The cross-section areas of each bar A, C, and D are calculated as $A_A = \dfrac{\pi d_A^2}{4}$, $A_C = \dfrac{\pi d_C^2}{4}$, and $A_D = \dfrac{\pi d_D^2}{4}$, respectively.

The cross-section areas are calculated in MATLAB with:

```
d_A=d;
d_C=d;
d_D=1.5*d;

A_A=pi*d_A*d_A/4;
A_C=pi*d_C*d_C/4;
A_D=pi*d_D*d_D/4;
fprintf('A_A = %s \n',char(A_A))
fprintf('A_C = %s \n',char(A_C))
fprintf('A_D = %s \n',char(A_D))
fprintf('\n')
```

The axial stresses on bars A, C, D, namely σ_A, σ_B, and σ_C, are calculated using

$$\sigma_A = \sigma_{34} = \frac{T_{34}}{A_A} = \frac{F_2}{A_A} = \frac{4 F_2}{\pi d_A^2},$$

$$\sigma_C = \sigma_{23} = \frac{T_{13}}{A_C} = \frac{F_2 - F_1}{A_C} = \frac{4 (F_2 - F_1)}{\pi d_C^2},$$

$$\sigma_D = \sigma_{12} = \frac{T_{13}}{A_D} = \frac{F_2 - F_1}{A_D} = \frac{4 (F_2 - F_1)}{\pi d_D^2}. \qquad (1.75)$$

The axial stresses are calculated in MATLAB with:

```
sigma_A=T_34/A_A;
sigma_C=T_13/A_C;
```

```
sigma_D=T_13/A_D;
fprintf('sigma_A = %s \n',char(sigma_A))
fprintf('sigma_C = %s \n',char(sigma_C))
fprintf('sigma_D = %s \n',char(sigma_D))
fprintf('\n')
```

The axial displacement of bar D is

$$\delta_{12} = \frac{T_{13}l_D}{A_D E} = \frac{(F_2 - F_1)l_D}{A_D E}.$$

The axial displacement of bar C is

$$\delta_{23} = \frac{T_{13}l_C}{A_C E} = \frac{(F_2 - F_1)l_C}{A_C E}.$$

The axial displacement of bars D and C is

$$\delta_{13} = \delta_{12} + \delta_{23} = \frac{(F_2 - F_1)l_D}{A_D E} + \frac{(F_2 - F_1)l_C}{A_C E} = \frac{F_2 - F_1}{E}\left(\frac{l_D}{A_D} + \frac{l_C}{A_C}\right).$$

The axial displacement of bar D is

$$\delta_{34} = \frac{T_{34}l_A}{A_A E} = \frac{F_2 l_A}{A_A E}.$$

The total axial displacement of the system is

$$\delta_{14} = \delta_{13} + \delta_{34} = \frac{F_2 - F_1}{E}\left(\frac{l_D}{A_D} + \frac{l_C}{A_C}\right) + \frac{F_2 l_A}{A_A E}. \tag{1.76}$$

The axial displacement of the bars and system are calculated in MATLAB with:

```
delta_12=T_13*l_D/(A_D*E);
delta_23=T_13*l_C/(A_C*E);
delta_13=delta_12+delta_23;
delta_34=T_34*l_A/(A_A*E);
delta_14=delta_13+delta_34;
fprintf('delta_12 = %s \n',char(delta_12))
fprintf('delta_23 = %s \n',char(delta_23))
fprintf('delta_13 = %s \n',char(delta_13))
fprintf('delta_34 = %s \n',char(delta_34))
fprintf('delta_41 = %s \n\n',char(delta_14))
```

The numerical data is introduced using the following code:

```
%numerical application
lists={F_1,F_2,d,l_A,l_C,l_D, E};
listn={192000,64000,45,170,144.5,255,21*10^4};
```

To numerically calculate the axial stress and displacements, the symbolic variables are substituted with the numerical data:

```
F1n=subs(F_1,lists,listn);
F2n=subs(F_2,lists,listn);
T_13n=subs(T_13,lists,listn);
T_34n=subs(T_34,lists,listn);
fprintf('F1 = %s(N)\n', char(F1n))
fprintf('F2 = %s (N)\n', char(F2n))
fprintf('T_13 = %s (N)\n', char(T_13n))
fprintf('T_34 = %s (N)\n', char(T_34n))
fprintf('\n')
```

The numerical values for the cross-section of each bar are calculated in MATLAB with:

```
d_An=subs(d_A,lists,listn);
d_Cn=subs(d_C,lists,listn);
d_Dn=subs(d_D,lists,listn);
A_An=subs(A_A,lists,listn);
A_Cn=subs(A_C,lists,listn);
A_Dn=subs(A_D,lists,listn);
fprintf('d_A = %s (mm)\n',char(d_An))
fprintf('d_C = %s (mm)\n',char(d_Cn))
fprintf('d_D = %s (mm)\n',char(d_Dn))
fprintf('A_A = %s (mm^2)\n',eval(char(A_An)))
fprintf('A_C = %s (mm^2)\n',eval(char(A_Cn)))
fprintf('A_D = %s (mm^2)\n',eval(char(A_Dn)))
fprintf('\n')
```

The numerical values of each bar diameter are expressed in MATLAB by:

```
l_An=subs(l_A,lists,listn);
l_Cn=subs(l_C,lists,listn);
l_Dn=subs(l_D,lists,listn);
fprintf('l_A = %s (mm)\n',char(l_An))
```

```
fprintf('l_C = %s (mm)\n',char(l_Cn))
fprintf('l_D = %s (mm)\n',char(l_Dn))
fprintf('\n')
```

The numerical values of the axial stresses are calculated in MATLAB with:

```
sigmA_An=subs(sigma_A,lists,listn);
sigmA_Cn=subs(sigma_C,lists,listn);
sigmA_Dn=subs(sigma_D,lists,listn);
fprintf('sigmA_A = %s (MPa)\n',eval(char(sigmA_An)))
fprintf('sigmA_C = %s (MPa)\n',eval(char(sigmA_Cn)))
fprintf('sigmA_D = %s (MPa)\n',eval(char(sigmA_Dn)))
fprintf('\n')
```

and the results are:

```
sigmA_A = 4.024066e+01 (MPa)
sigmA_C = -8.048131e+01 (MPa)
sigmA_D = -3.576947e+01 (MPa)
```

The numerical values of the axial displacement of the bars and system are calculated in MATLAB with:

```
delta_12n=subs(delta_12,lists,listn);
delta_23n=subs(delta_23,lists,listn);
delta_13n=subs(delta_13,lists,listn);
delta_34n=subs(delta_34,lists,listn);
delta_14n=subs(delta_14,lists,listn);
delta_24n=delta_34n+delta_23n;
fprintf('delta_12 = %s (mm)\n',eval(char(delta_12n)))
fprintf('delta_14 = %s (mm)\n',eval(char(delta_14n)))
fprintf('delta_23 = %s (mm)\n',eval(char(delta_23n)))
fprintf('delta_34 = %s (mm)\n',eval(char(delta_34n)))
fprintf('delta_24 = %s (mm)\n',eval(char(delta_24n)))
fprintf('\n')
```

and the results are:

```
delta_12 = -4.343436e-02 (mm)
delta_14 = -6.623740e-02 (mm)
delta_23 = -5.537881e-02 (mm)
delta_34 = 3.257577e-02 (mm)
delta_24 = -2.280304e-02 (mm)
```

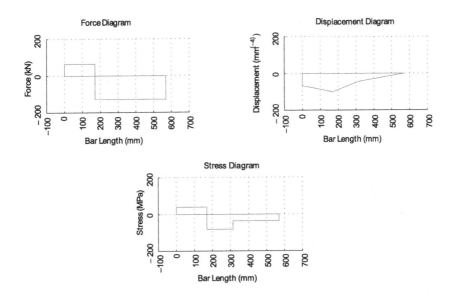

Figure 1.15 Force, stress and displacement diagrams.

The force, stress and displacement diagrams obtained using the following MATLAB code are shown in Fig. 1.15.

```
%Force Diagram
subplot(1,3,1)
ForceDdraw(T_13n,T_34n,l_An,l_Cn,l_Dn);
%Stress Diagram
subplot(1,3,2)
StressDdraw(sigmA_An,sigmA_Cn,...
    sigmA_Dn,l_An,l_Cn,l_Dn);
%Displacement Diagram
subplot(1,3,3)
DisplacementDdraw(delta_14n,delta_13n,...
    delta_12n,l_An,l_Cn,l_Dn);
```

The MATLAB function *ForceDdraw* is:

```
function []=ForceDdraw(F1A,F2A,l1,l2,l3,l4)
xlabel('Force (kN)'), ylabel('Bar Length (mm)')
title('Force Diagram')
hold on; grid on
axis equal; axis on
```

```
axis([-200 200 -100 700])
hold on;
scale_F=1000;
hold on
l=line([0 0],[0 (l1+l2+l3)]);
l=line([0 F1A/scale_F],[(l1+l2+l3) (l1+l2+l3)]);
l=line([F1A/scale_F F1A/scale_F],...
    [(l1+l2+l3) (l1+l2+l3)-(l3)]);
l=line([F1A/scale_F F1A/scale_F],...
    [(l1+l2+l3)-l3 (l1+l2+l3)-(l3)]);
l=line([F1A/scale_F F1A/scale_F],...
    [(l1+l2+l3)-l3 (l1+l2+l3)-(l2+l3)]);
l=line([F1A/scale_F F2A/scale_F],...
    [(l1+l2+l3)-(l2+l3) (l1+l2+l3)-(l2+l3)]);
l=line([F2A/scale_F F2A/scale_F],...
    [0 (l1+l2+l3)-(l2+l3)]);
l=line([F2A/scale_F 0],[0 0]);
set(l,'Color','blue');
end
```

The MATLAB function *StressDdraw* is:

```
function []=StressDdraw(sigma1,sigma2,sigma3,l1,l2,l3)
xlabel('Stress (MPa)'), ylabel('Bar Length (mm)')
title('Stress Diagram')
hold on; grid on
axis equal; axis on
axis([-200 200 -100 700])
hold on;
scale_S=1;
hold on
l=line([0 0],[0 (l1+l2+l3)]);
l=line([0 sigma3],[(l1+l2+l3) (l1+l2+l3)]);
l=line([sigma3/scale_S sigma3/scale_S],...
    [(l1+l2+l3) (l1+l2+l3)-(l3)]);
l=line([sigma3/scale_S sigma2/scale_S],...
    [(l1+l2+l3)-(l3) (l1+l2+l3)-(l3)]);
l=line([sigma2/scale_S sigma2/scale_S],...
    [(l1+l2+l3)-(l3) (l1+l2+l3)-(l2+l3)]);
l=line([sigma2/scale_S sigma1/scale_S],...
```

```
    [(l1+l2+l3)-(l2+l3) (l1+l2+l3)-(l2+l3)]);
l=line([sigma1/scale_S sigma1/scale_S],...
    [0 (l1+l2+l3)-(l2+l3)]);
l=line([sigma1/scale_S 0],[0 0]);
set(l,'Color','blue');
end
```

The MATLAB function *DisplacementDdraw* is:

```
function []=DisplacementDdraw(delta1,delta2,delta3,l1,l2,l3)
xlabel('Displacement (mm^{(-4)})'), ylabel('Bar Length (mm)')
title('Displacement Diagram')
hold on; grid on
axis equal; axis on
axis([-200 200 -100 700])
hold on;
scale_D=10^(-3);
hold on
l=line([0 0],[0 (l1+l2+l3)]);
l=line([0 delta3/scale_D],...
    [(l1+l2+l3) (l1+l2+l3)-(l3)]);
l=line([delta3/scale_D delta2/scale_D],...
    [(l1+l2+l3)-(l3) (l1+l2+l3)-(l2+l3)]);
l=line([delta2/scale_D delta1/scale_D],...
    [(l1+l2+l3)-(l2+l3) (l1+l2+l3)-(l1+l2+l3)]);
l=line([delta1/scale_D 0],[0 0]);
set(l,'Color','blue');
end
```

Example 1.2. A rigid plate B is fastened to bar A and bar C, which is also fastened to bar D. A second rigid plate E is fastened to bars D and F as shown in Fig. 1.16. The ends of bars A and F are fastened to rigid supports 1 and 2. Bars A, C, D, and E have lengths l_A, l_C, l_D, l_F and diameters $d_A = d_C = d$, $d_D = d_F = 1.5d$, respectively. Load F_1 is applied to the rigid plate B (distributed uniformly around the circumference of the rigid plate B), and load F_2 is applied to the rigid plate E (distributed uniformly around the circumference of the rigid plate E) as shown in Fig. 1.16. Determine the axial stresses in bars A, C, D, and E, deformations of the bars, total deformation of the system, and reactions at ends 1 and 2.

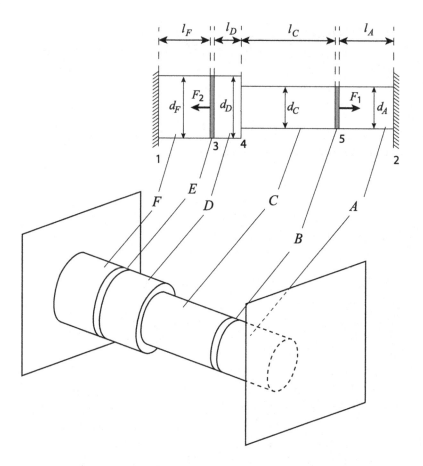

Figure 1.16 Axial bar under loading.

For the numerical application use: $F_1 = 24\,000$ N, $F_2 = 48\,000$ N, $d = 15$ mm, $l_A = 65$ mm, $l_C = 227.5$ mm, $l_D = 32.5$ mm, $l_F = 48.75$ mm, and $E = 21\,(10^4)$ MPa.

Solution

The system is divided into components, and the internal forces are calculated passing sections through each component. Free-body diagrams are developed for portions of the bar.

The free-body diagram of the system, namely bar F, plate E, bar D, bar C, plate B, and bar A, contains only two unknown forces, that is, reactions R_1 and R_2 at the rigid supports as shown in Fig. 1.17A. One can

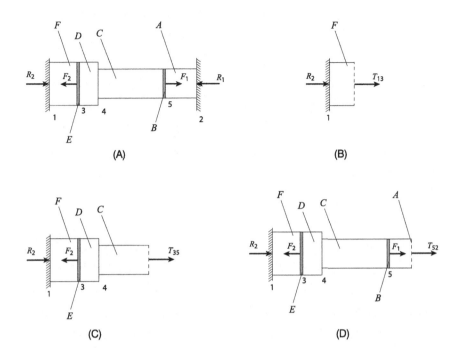

Figure 1.17 Free-body diagram.

write

$$\sum F_y^{A,B,C,D,E,F} = 0 \Leftrightarrow -R_2 + F_2 - F_1 + R_1 = 0 \Leftrightarrow R_1 - R_2 + F_2 - F_1 = 0.$$

(1.77)

The internal forces are calculated for each portion of the bar. For the interval 1–3, the free-body diagram of the bar shown in Fig. 1.17B gives

$$\sum F_x = 0 \Leftrightarrow T_{13} + R_2 = 0 \Leftrightarrow T_{13} = -R_2.$$ (1.78)

For the interval 3–5, the free-body diagram shown in Fig. 1.17C gives

$$\sum F_x = 0 \Leftrightarrow T_{35} + R_2 - F_2 = 0 \Leftrightarrow T_{35} = F_2 - R_2.$$ (1.79)

For the interval 5–2, the free body diagram shown in Fig. 1.17D gives

$$\sum F_x = 0 \Leftrightarrow T_{52} + R_2 - F_2 + F_1 = 0 \Leftrightarrow T_{52} = F_2 - R_2 - F_1.$$ (1.80)

The MATLAB program starts with:

```
clear all; clc; close all
syms F_1 F_2 T_13 T_35 T_52 R_1 R_2 R_2_
syms l_A d_A l_C d_C l_D d_D d E l_F
```

The equilibrium equations are written and displayed in MATLAB with:

```
sumFOy = R_1 - F_1 + F_2 - R_2;
sumFOy1 = T_13 + R_2;
sumFOy2 = T_35 + R_2 - F_2;
sumFOy3 = T_52 + R_2 + F_1 - F_2;
fprintf('F_1 = %s \n',char(F_2))
fprintf('F_2 = %s \n',char(F_1))
fprintf('\n')
fprintf('sumFOy = %s \n',char(sumFOy))
fprintf('sumFOy1 = %s \n',char(sumFOy1))
fprintf('sumFOy2 = %s \n',char(sumFOy2))
fprintf('sumFOy3 = %s \n\n',char(sumFOy3))
```

Force calculation in each section is performed in MATLAB with:

```
soll1=solve(sumFOy1,T_13);
T_13 = soll1;
fprintf('T_13 = %s \n',char(T_13))
soll2=solve(sumFOy2,T_35);
T_35 = soll2;
fprintf('T_35 = %s \n',char(T_35))
soll3=solve(sumFOy3,T_52);
T_52 = soll3;
fprintf('T_52 = %s \n\n',char(T_52))
```

The cross-section areas of each bar A, C, D, and F are calculated using $A_A = \dfrac{\pi d_A^2}{4}$, $A_C = \dfrac{\pi d_C^2}{4}$, $A_D = \dfrac{\pi d_D^2}{4}$, $A_F = \dfrac{\pi d_F^2}{4}$, respectively. The cross-section areas are calculated and printed in MATLAB with:

```
d_A=d;
d_C=d;
d_D=1.5*d;
d_F=1.5*d;
A_A=pi*d_A*d_A/4;
```

```
A_C=pi*d_C*d_C/4;
A_D=pi*d_D*d_D/4;
A_F=pi*d_F*d_F/4;
fprintf('A_A = %s \n',char(A_A))
fprintf('A_C = %s \n',char(A_C))
fprintf('A_D = %s \n',char(A_D))
fprintf('A_F = %s \n\n',char(A_F))
```

The axial stresses σ_A, σ_C, σ_D, and σ_F are calculated using

$$
\begin{aligned}
\sigma_F &= \frac{F_{1A}}{A_F} = \frac{-R_2}{A_F} = -\frac{4R_2}{\pi d_F^2}, \\
\sigma_D &= \frac{F_{2A}}{A_D} = \frac{F_2 - R_2}{A_D} = \frac{4(F_2 - R_2)}{\pi d_D^2}, \\
\sigma_C &= \frac{F_{2A}}{A_C} = \frac{F_2 - R_2}{A_C} = \frac{4(F_2 - R_2)}{\pi d_C^2}, \\
\sigma_A &= \frac{F_{3A}}{A_A} = \frac{F_2 - R_2 - F_1}{A_A} = \frac{4(F_2 - R_2 - F_1)}{\pi d_A^2}.
\end{aligned}
\tag{1.81}
$$

The MATLAB computation for the axial stresses is given by:

```
sigma_A=T_52/A_A;
sigma_C=T_35/A_C;
sigma_D=T_35/A_D;
sigma_F=T_13/A_F;
fprintf('sigma_A = %s \n',char(sigma_A))
fprintf('sigma_C = %s \n',char(sigma_C))
fprintf('sigma_D = %s \n',char(sigma_D))
fprintf('sigma_F = %s \n\n',char(sigma_F))
```

The axial displacements of bars A, C, D, and F are

$$
\begin{aligned}
\delta_1 &= 0, \\
\delta_2 &= 0, \\
\delta_{3-1} &= \delta_3 = \frac{F_{1A} l_F}{A_F E} = -\frac{R_2 l_F}{A_F E}, \\
\delta_{3-4} &= \frac{F_{2A} l_D}{A_D E} = \frac{(F_2 - R_2) l_D}{A_D E}, \\
\delta_{4-1} &= \delta_4 = \delta_3 + \delta_{3-4} = -\frac{R_2 l_F}{A_F E} + \frac{(F_2 - R_2) l_D}{A_D E},
\end{aligned}
$$

$$\delta_{4-5} = \frac{F_{2A}l_C}{A_C E} = \frac{(F_2 - R_2)\,l_C}{A_C E},$$

$$\delta_{5-1} = \delta_5 = \delta_4 + \delta_{4-5} = -\frac{R_2 l_F}{A_F E} + \frac{(F_2 - R_2)\,l_D}{A_D E} + \frac{(F_2 - R_2)\,l_C}{A_C E},$$

$$\delta_{5-2} = \frac{F_{3A}l_A}{A_A E} = \frac{(F_2 - R_2 - F_1)\,l_A}{A_A E}. \tag{1.82}$$

The displacements of the bars are calculated in MATLAB with:

```
delta_1=0;
delta_2=0;
delta_31=T_13*1_F/(A_F*E);
delta_34=T_35*1_D/(A_D*E);
delta_41=delta_31+delta_34;
delta_45=T_35*1_C/(A_C*E);
delta_51=delta_41+delta_45;
delta_52=T_52*1_A/(A_A*E);
fprintf('delta_1 = %f \n',delta_1)
fprintf('delta_2 = %f \n',delta_2)
fprintf('delta_31 = %s \n',char(delta_31))
fprintf('delta_41 = %s \n',char(delta_41))
fprintf('delta_51 = %s \n',char(delta_51))
fprintf('delta_34 = %s \n',char(delta_31))
fprintf('delta_45 = %s \n',char(delta_45))
fprintf('delta_52 = %s \n\n',char(delta_52))
```

Since the ends A and F of the bars are fastened to rigid supports, there is no displacement at the bar ends, that is, $\delta_{total} = 0$. Therefore, one can write

$$\begin{aligned}
\delta_{total} &= \delta_{2-1} = \delta_{1-3} + \delta_{3-4} + \delta_{4-5} + \delta_{5-2} \\
&= -\frac{R_2 l_F}{A_F E} + \frac{(F_2 - R_2)\,l_D}{A_D E} \\
&+ \frac{(F_2 - R_2)\,l_C}{A_C E} + \frac{(F_2 - R_2 - F_1)\,l_A}{A_A E} = 0.
\end{aligned} \tag{1.83}$$

In MATLAB this is achieved by:

```
delta_21=delta_31+delta_34+delta_45+delta_52;
delta_tot=delta_21;
fprintf('delta_total = %s \n',char(delta_tot))
```

It results in

$$-\frac{R_2 l_F}{A_F E} + \frac{F_2 l_D}{A_D E} - \frac{R_2 l_D}{A_D E} + \frac{F_2 l_C}{A_C E} - \frac{R_2 l_C}{A_C E} + \frac{(F_2 - F_1) l_A}{A_A E} - \frac{R_2 l_A}{A_A E} = 0$$

$$\Leftrightarrow \quad \frac{R_2 l_F}{A_F E} + \frac{R_2 l_D}{A_D E} + \frac{R_2 l_C}{A_C E} + \frac{R_2 l_A}{A_A E} = \frac{F_2 l_D}{A_D E} + \frac{F_2 l_C}{A_C E} + \frac{(F_2 - F_1) l_A}{A_A E}$$

$$\Leftrightarrow \quad R_2 \left(\frac{l_F}{A_F E} + \frac{l_D}{A_D E} + \frac{l_C}{A_C E} + \frac{l_A}{A_A E} \right)$$

$$= F_2 \frac{l_D}{A_D E} + F_2 \frac{l_C}{A_C E} + (F_2 - F_1) \frac{l_A}{A_A E}$$

$$\Leftrightarrow \quad R_2 = \frac{F_2 \dfrac{l_D}{A_D E} + F_2 \dfrac{l_C}{A_C E} + (F_2 - F_1) \dfrac{l_A}{A_A E}}{\dfrac{l_F}{A_F E} + \dfrac{l_D}{A_D E} + \dfrac{l_C}{A_C E} + \dfrac{l_A}{A_A E}}. \tag{1.84}$$

Using Eq. (1.77), one can write

$$R_1 = R_2 - F_2 + F_1 = 0. \tag{1.85}$$

From Eqs. (1.85) and (1.84), one can write

$$R_1 = \frac{F_2 \dfrac{l_D}{A_D E} + F_2 \dfrac{l_C}{A_C E} + (F_2 - F_1) \dfrac{l_A}{A_A E}}{\dfrac{l_F}{A_F E} + \dfrac{l_D}{A_D E} + \dfrac{l_C}{A_C E} + \dfrac{l_A}{A_A E}} - F_2 + F_1 = 0. \tag{1.86}$$

Reactions R_1 and R_2 at the rigid supports are calculated in MATLAB using the code:

```
solla=solve(delta_tot,R_2);
R_2new = solla;
fprintf('R_2 = %s \n',char(R_2new))
sollb=solve(sumFOy,R_1);
R_1 = subs(sollb,R_2,R_2new);
fprintf('R_1 = %s \n\n',char(R_1))
```

Using the newly calculated reactions R_1 and R_2, the axial displacements can be expressed in MATLAB with:

```
delta_31=subs(delta_31,R_2,R_2new);
delta_34=subs(delta_34,R_2,R_2new);
```

```
delta_41=delta_31+delta_34;
delta_45=subs(delta_45,R_2,R_2new);
delta_51=delta_41+delta_45;
delta_52=subs(delta_52,R_2,R_2new);
fprintf('delta_1 = %f \n',delta_1)
fprintf('delta_2 = %f \n',delta_2)
fprintf('delta_31 = %s \n',char(delta_31))
fprintf('delta_41 = %s \n',char(delta_41))
fprintf('delta_51 = %s \n',char(delta_51))
fprintf('delta_34 = %s \n',char(delta_34))
fprintf('delta_45 = %s \n',char(delta_45))
fprintf('delta_52 = %s \n\n',char(delta_52))
```

The input data are given in MATLAB as:

```
%numerical application
lists={F_1,F_2,d,l_A,l_C,l_D,l_F,E};
listn={24000,48000,15,65,3.5*65,0.5*65,0.75*65,21*10^4};
```

The numerical values for the displacements are calculated in MATLAB using:

```
delta_31_=subs(delta_31,lists,listn);
delta_41_=subs(delta_41,lists,listn);
delta_51_=subs(delta_51,lists,listn);
delta_34_=subs(delta_34,lists,listn);
delta_45_=subs(delta_45,lists,listn);
delta_52_=subs(delta_52,lists,listn);
fprintf('delta_1_ = %f (mm)\n',delta_1)
fprintf('delta_2_ = %f (mm)\n',delta_2)
fprintf('delta_31_ = %s (mm)\n',eval(char(delta_31_)))
fprintf('delta_41_ = %s (mm)\n',eval(char(delta_41_)))
fprintf('delta_51_ = %s (mm)\n',eval(char(delta_51_)))
fprintf('delta_34_ = %s (mm)\n',eval(char(delta_34_)))
fprintf('delta_45_ = %s (mm)\n',eval(char(delta_45_)))
fprintf('delta_52_ = %s (mm)\n\n',eval(char(delta_52_)))
```

The results are:

```
delta_1_ = 0.000000 (mm)
delta_2_ = 0.000000 (mm)
delta_31_ = -2.340528e-02 (mm)
```

```
delta_41_ = -2.032564e-02 (mm)
delta_51_ = 2.817873e-02 (mm)
delta_34_ = 3.079642e-03 (mm)
delta_45_ = 4.850436e-02 (mm)
delta_52_ = -2.817873e-02 (mm)
```

The numerical values of reactions R_1 and R_2 are obtained and printed in MATLAB with:

```
R_2_=subs(R_2new,lists,listn);
R_1_=subs(R_1,lists,listn);
fprintf('R_1_ = %s (N)\n',eval(char(R_1_)))
fprintf('R_2_ = %s (N)\n\n',eval(char(R_2_)))
```

The values of the reactions are:

```
R_1_ = 1.608791e+04 (N)
R_2_ = 4.008791e+04 (N)
```

Using the numerical values of reactions R_1 and R_2, the values of the axial stresses can be obtained in MATLAB with:

```
sigma_A_=eval(subs(sigma_A,[lists,R_2],...
    [listn,R_2_]));
sigma_C_=eval(subs(sigma_C,[lists,R_2],...
    [listn,R_2_]));
sigma_D_=eval(subs(sigma_D,[lists,R_2],...
    [listn,R_2_]));
sigma_F_=eval(subs(sigma_F,[lists,R_2],...
    [listn,char(R_2_)]));
fprintf('sigma_A_ = %s (MPa)\n',sigma_A_)
fprintf('sigma_C_ = %s (MPa)\n',sigma_C_)
fprintf('sigma_D_ = %s (MPa)\n',sigma_D_)
fprintf('sigma_F_ = %s (MPa)\n\n',sigma_F_)
```

The numerical values for the axial stresses σ_A, σ_C, σ_D, and σ_F are:

```
sigma_A_ = -9.103896e+01 (MPa)
sigma_C_ = 4.477326e+01 (MPa)
sigma_D_ = 1.989923e+01 (MPa)
sigma_F_ = -1.008227e+02 (MPa)
```

The force, stress and displacement diagrams obtained using the following MATLAB code are shown in Fig. 1.18.

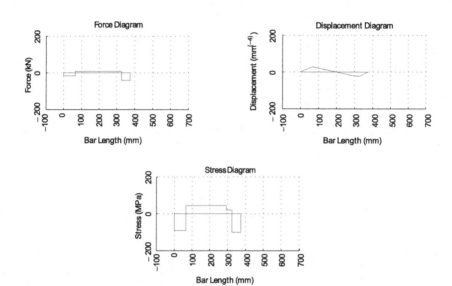

Figure 1.18 Force, stress and displacement diagrams.

```
T_13_=subs(T_13,[lists,R_2],[listn,eval(R_2_)]);
T_35_=subs(T_35,[lists,R_2],[listn,R_2_]);
T_52_=subs(T_52,[lists,R_2],[listn,R_2_]);
l_A_=subs(l_A,lists,listn);l_C_=subs(l_C,lists,listn);
l_D_=subs(l_D,lists,listn);l_F_=subs(l_F,lists,listn);
%Force Diagram
subplot(1,3,1)
ForceDadraw(T_13_,T_35_,T_52_,l_A_,l_C_,l_D_,l_F_);
%Stress Diagram
subplot(1,3,2)
StressDadraw(sigma_A_,sigma_C_,sigma_D_,sigma_F_,...
    l_A_,l_C_,l_D_,l_F_);
%Displacement Diagram
subplot(1,3,3)
DisplacementDadraw(delta_51_,delta_41_,delta_31_,...
    l_A_,l_C_,l_D_,l_F_);
```

The MATLAB function *ForceDadraw* is:

```
function []=ForceDadraw(F1A,F2A,F3A,l1,l2,l3,l4)
xlabel('Force (kN)'), ylabel('Bar Length (mm)')
title('Force Diagram')
```

```
hold on; grid on
axis equal; axis on
axis([-200 200 -100 700])
hold on;
scale_F=1000;
hold on
l=line([0 0],[0 (l1+l2+l3+l4)]);
l=line([0 F1A/scale_F],...
    [(l1+l2+l3+l4) (l1+l2+l3+l4)]);
l=line([F1A/scale_F F1A/scale_F],...
    [(l1+l2+l3+l4) (l1+l2+l3+l4)-(l4)]);
l=line([F1A/scale_F F2A/scale_F],...
    [(l1+l2+l3+l4)-l4 (l1+l2+l3+l4)-(l4)]);
l=line([F2A/scale_F F2A/scale_F],...
    [(l1+l2+l3+l4)-l4 (l1+l2+l3+l4)-(l3+l4)]);
l=line([F2A/scale_F F2A/scale_F],...
    [(l1+l2+l3+l4)-(l3+l4) (l1+l2+l3+l4)-(l3+l4)]);
l=line([F2A/scale_F F2A/scale_F],...
    [(l1+l2+l3+l4)-(l3+l4) (l1+l2+l3+l4)-(l2+l3+l4)]);
l=line([F2A/scale_F F3A/scale_F],...
    [(l1+l2+l3+l4)-(l2+l3+l4) (l1+l2+l3+l4)-(l2+l3+l4)]);
l=line([F3A/scale_F F3A/scale_F],...
    [0 (l1+l2+l3+l4)-(l2+l3+l4)]);
l=line([F3A/scale_F 0],[0 0]);
set(l,'Color','blue');
end
```

The MATLAB function *StressDadraw* is:

```
function []=StressDadraw(sigma1,sigma2,...
    sigma3,sigma4,l1,l2,l3,l4)
xlabel('Stress (MPa)'), ylabel('Bar Length (mm)')
title('Stress Diagram')
hold on; grid on
axis equal; axis on
axis([-200 200 -100 700])
hold on;
scale_S=1;
hold on
l=line([0 0],[0 (l1+l2+l3+l4)]);
```

```
l=line([0 sigma4/scale_S],...
    [(l1+l2+l3+l4) (l1+l2+l3+l4)]);
l=line([sigma4/scale_S sigma4/scale_S],...
    [(l1+l2+l3+l4) (l1+l2+l3+l4)-(l4)]);
l=line([sigma4/scale_S sigma3/scale_S],...
    [(l1+l2+l3+l4)-l4 (l1+l2+l3+l4)-(l4)]);
l=line([sigma3/scale_S sigma3/scale_S],...
    [(l1+l2+l3+l4)-l4 (l1+l2+l3+l4)-(l3+l4)]);
l=line([sigma3/scale_S sigma2/scale_S],...
    [(l1+l2+l3+l4)-(l3+l4) (l1+l2+l3+l4)-(l3+l4)]);
l=line([sigma2/scale_S sigma2/scale_S],...
    [(l1+l2+l3+l4)-(l3+l4) (l1+l2+l3+l4)-(l2+l3+l4)]);
l=line([sigma2/scale_S sigma1/scale_S],...
    [(l1+l2+l3+l4)-(l2+l3+l4) (l1+l2+l3+l4)-(l2+l3+l4)]);
l=line([sigma1/scale_S sigma1/scale_S],...
    [0 (l1+l2+l3+l4)-(l2+l3+l4)]);
l=line([sigma1/scale_S 0],[0 0]);
set(l,'Color','blue');
end
```

The MATLAB function *DisplacementDadraw* is:

```
function []=DisplacementDadraw(delta1,delta2,...
    delta3,l1,l2,l3,l4)
xlabel('Displacement (mm^{(-4)})'), ylabel('Bar Length (mm)')
title('Displacement Diagram')
hold on; grid on
axis equal; axis on
axis([-200 200 -100 700])
hold on;
scale_D=10^(-3);
hold on
l=line([0 0],[0 (l1+l2+l3+l4)]);
l=line([0 delta3/scale_D],...
    [(l1+l2+l3+l4) (l1+l2+l3+l4)-(l4)]);
l=line([delta3/scale_D delta2/scale_D],...
    [(l1+l2+l3+l4)-(l4) (l1+l2+l3+l4)-(l3+l4)]);
l=line([delta2/scale_D delta1/scale_D],...
    [(l1+l2+l3+l4)-(l3+l4) (l1+l2+l3+l4)-(l2+l3+l4)]);
l=line([delta1/scale_D 0],...
```

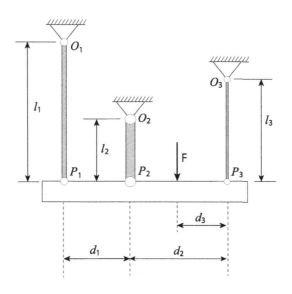

Figure 1.19 System of bars under loading.

```
    [(l1+l2+l3+l4)-(l2+l3+l4) (l1+l2+l3+l4)-(l1+l2+l3+l4)]);
set(1,'Color','blue');
end
```

Example 1.3. The vertical load F shown in Fig. 1.19 is supported by three different bars denoted by O_1P_1, O_2P_2, and O_3P_3. Bar O_1P_1 has length l_1, radius r_1 and modulus of elasticity E_1; bar O_2P_2 has length l_2, radius r_2, and modulus of elasticity E_2; and bar O_3P_3 has length l_3, radius r_3, and modulus of elasticity E_3. The distance between points P_1 and P_2 is $d_{P_1P_2} = d_1$, the distance between points P_2 and P_3 is $d_{P_2P_3} = d_2$, and load F is acting at a distance d_3 from point P_3 as shown in Fig. 1.19. Determine the load, axial stresses and displacements of the bars when the externally applied load acts as shown.

Solution

The free-body diagram of the system in Fig. 1.20 contains tree un-known forces, reactions R_1, R_2, and R_3. Resolving forces vertically, one can obtain

$$\sum F_{(Oy)} = 0 \Leftrightarrow R_1 + R_2 - F + R_3 = 0. \tag{1.87}$$

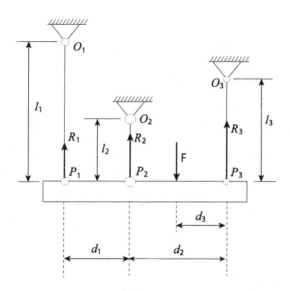

Figure 1.20 Free-body diagram of the system of bars under loading.

Taking moments about P_1 (the sum of moments about point P_1 is zero), one can obtain

$$\sum M_{P_1} = 0 \Leftrightarrow R_3 (d_1 + d_2) - F (d_1 + d_2 - d_3) + R_2 d_1 = 0. \qquad (1.88)$$

The equilibrium equations in MATLAB, i.e., sums of forces and moments, are written and displayed using

```
sumFOy = R_1 + R_2 - F + R_3;
sumM_P1 = R_3*(d1+d2)...
    - F*(d1+d2-d3) + R_2*d1;
fprintf('sumFOy = %s \n',...
    char(sumFOy))
fprintf('\n')
fprintf('sumM_P1 = %s \n',...
    char(sumM_P1))
fprintf('\n')
```

Another equation is needed to determine reactions R_1, R_2, and R_3 (Fig. 1.20). Using a geometric approach (Fig. 1.21), one can write

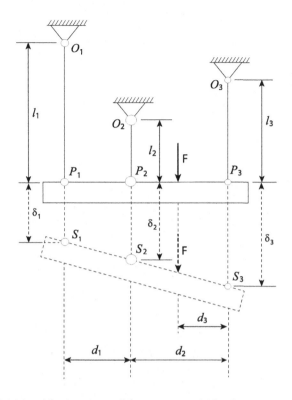

Figure 1.21 Initial and final position of the system under loading.

$$\frac{\delta_2 - \delta_1}{\delta_3 - \delta_1} = \frac{d_1}{d_1 + d_2}$$

$$\Leftrightarrow \frac{\dfrac{R_2 l_{O_2 P_2}}{A_{O_2 P_2} E_{O_2 P_2}} - \dfrac{R_1 l_{O_1 P_1}}{A_{O_1 P_1} E_{O_1 P_1}}}{\dfrac{R_3 l_{O_3 P_3}}{A_{O_3 P_3} E_{O_3 P_3}} - \dfrac{R_1 l_{O_1 P_1}}{A_{O_1 P_1} E_{O_1 P_1}}} = \frac{d_1}{d_1 + d_2}$$

$$\Leftrightarrow \frac{\dfrac{R_2 l_2}{A_2 E_2} - \dfrac{R_1 l_1}{A_1 E_1}}{\dfrac{R_3 l_3}{A_3 E_3} - \dfrac{R_1 l_1}{A_1 E_1}} = \frac{d_1}{d_1 + d_2}$$

$$\Leftrightarrow \frac{R_2 l_2}{A_2 E_2} - \frac{R_1 l_1}{A_1 E_1} = \frac{d_1}{d_1 + d_2} \left(\frac{R_3 l_3}{A_3 E_3} - \frac{R_1 l_1}{A_1 E_1} \right)$$

$$\Leftrightarrow \frac{R_2 l_2}{r_2^2 E_2} - \frac{R_1 l_1}{r_1^2 E_1} = \frac{d_1}{d_1 + d_2} \left(\frac{R_3 l_3}{r_3^2 E_3} - \frac{R_1 l_1}{r_1^2 E_1} \right), \tag{1.89}$$

where $A_{O_1P_1} = A_1 = \pi r_1^2$, $A_{O_2P_2} = A_2 = \pi r_2^2$, and $A_{O_3P_3} = A_3 = \pi r_3^2$ represent the cross-section areas of bars O_1P_1, O_2P_2, and O_3P_3, respectively. The equation is written in MATLAB using:

```
A_1=pi*r1^2;
A_2=pi*r2^2;
A_3=pi*r3^2;
delta_1=(R_1*l1)/(A_1*E1);
delta_2=(R_2*l2)/(A_2*E2);
delta_3=(R_3*l3)/(A_3*E3);

eq3 = (delta_2-delta_1)/(delta_3-delta_1)...
    - d1/(d1+d2);
fprintf('eq3 = %s \n',char(eq3))
fprintf('\n')
```

Using Eqs. (1.87), (1.88), and (1.89), one can write

$$\begin{cases} R_1 + R_2 - F + R_3 = 0, \\ R_3\left(d_1 + d_2\right) - F\left(d_1 + d_2 - d_3\right) + R_2 d_1 = 0, \\ \dfrac{R_2 l_2}{r_2^2 E_2} - \dfrac{R_1 l_1}{r_1^2 E_1} = \dfrac{d_1}{d_1 + d_2}\left(\dfrac{R_3 l_3}{r_3^2 E_3} - \dfrac{R_1 l_1}{r_1^2 E_1}\right). \end{cases} \qquad (1.90)$$

Solving Eq. (1.90), one can obtain

$$R_1 = \frac{E_1 F r_1^2 \left[E_2 d_1 l_3 r_2^2 \left(d_3 - d_2\right) + E_3 d_3 l_2 r_3^2 \left(d_1 + d_2\right)\right]}{E_1 E_2 l_3 d_1^2 r_1^2 r_2^2 + E_1 E_3 l_2 r_1^2 r_3^2 \left(d_1^2 + 2 d_1 d_2 + d_2^2\right) + E_2 E_3 l_1 d_2^2 r_2^2 r_3^2},$$

$$R_2 = \frac{E_2 F r_2^2 \left[E_1 d_1 l_3 r_1^2 \left(d_1 + d_2 - d_3\right) + E_3 d_2 d_3 l_1 r_3^2\right]}{E_1 E_2 l_3 d_1^2 r_1^2 r_2^2 + E_1 E_3 l_2 r_1^2 r_3^2 \left(d_1^2 + 2 d_1 d_2 + d_2^2\right) + E_2 E_3 l_1 d_2^2 r_2^2 r_3^2},$$

$$R_3 = \frac{E_3 F r_3^2 \left[E_1 d_1 l_2 r_1^2 \left(d_1 + 2 d_2 - d_3\right) + \left(E_1 d_2 l_2 r_1^2 + E_2 d_2 l_1 r_2^2\right)\left(d_2 - d_3\right)\right]}{E_1 E_2 l_3 d_1^2 r_1^2 r_2^2 + E_1 E_3 l_2 r_1^2 r_3^2 \left(d_1^2 + 2 d_1 d_2 + d_2^2\right) + E_2 E_3 l_1 d_2^2 r_2^2 r_3^2}.$$

$$(1.91)$$

The system of equations is solved in MATLAB using:

```
sol=solve(sumFOy,sumM_P1,eq3,...
    R_1,R_2,R_3);
R1 = simplify(sol.R_1);
R2 = simplify(sol.R_2);
```

```
R3 = simplify(sol.R_3);
fprintf('R1 = %s \n',char(R1))
fprintf('R2 = %s \n',char(R2))
fprintf('R3 = %s \n',char(R3))
fprintf('\n')
```

The axial stresses $\sigma_1 = \sigma_{O_1 P_1}$, $\sigma_2 = \sigma_{O_2 P_2}$, and $\sigma_3 = \sigma_{O_3 P_3}$ are calculated using

$$
\begin{aligned}
\sigma_3 &= \frac{R_3}{A_{O_3 P_3}} = \frac{R_3}{A_3} = \frac{R_3}{\pi r_3^2}, \\
\sigma_2 &= \frac{R_2}{A_{O_2 P_2}} = \frac{R_3}{A_2} = \frac{R_2}{\pi r_2^2}, \\
\sigma_1 &= \frac{R_1}{A_{O_1 P_1}} = \frac{R_3}{A_1} = \frac{R_1}{\pi r_1^2},
\end{aligned}
\tag{1.92}
$$

and in MATLAB with:

```
sigma_1=R1/A_1;
sigma_2=R2/A_2;
sigma_3=R3/A_3;
fprintf('sigma_1 = %s \n',char(sigma_1))
fprintf('sigma_2 = %s \n',char(sigma_2))
fprintf('sigma_3 = %s \n',char(sigma_3))
fprintf('\n')
fprintf('\n')
```

Considering the axial stresses σ_1, σ_2, and σ_3, one can obtain the displacements of the bars using

$$
\begin{aligned}
\delta_3 &= \frac{R_3 l_{O_3 P_3}}{A_{O_3 P_3} E_{O_3 P_3}} = \frac{R_3 l_3}{\pi r_3^2 E_3}, \\
\delta_2 &= \frac{R_2 l_{O_2 P_2}}{A_{O_2 P_2} E_{O_2 P_2}} = \frac{R_2 l_2}{\pi r_2^2 E_2}, \\
\delta_1 &= \frac{R_1 l_{O_1 P_1}}{A_{O_1 P_1} E_{O_1 P_1}} = \frac{R_1 l_1}{\pi r_1^2 E_1}.
\end{aligned}
\tag{1.93}
$$

Bar displacements are calculated in MATLAB with:

```
delta1=(R1*l1)/(A_1*E1);
delta2=(R2*l2)/(A_2*E2);
delta3=(R3*l3)/(A_3*E3);
fprintf('delta_1 = %s \n',char(simplify(delta1)))
```

```
fprintf('delta_2 = %s \n',char(delta2))
fprintf('delta_3 = %s \n',char(delta3))
fprintf('\n')
```

The input data are given in MATLAB with:

```
%numerical application
lists={F,r1,r2,r3,d1,d2,d3,l1,l2,l3,E1,E2,E3};
listn={32000,25,45,15,120,280,100,125,65,95,...
    21*10^4,21*10^4,21*10^4};
```

Reactions R_1, R_2, and R_3 are calculated and printed in MATLAB with:

```
R1 = subs(R1,lists,listn);
R2 = subs(R2,lists,listn);
R3 = subs(R3,lists,listn);
fprintf('R1 = %s (N)\n',eval(char(R1)))
fprintf('R2 = %s (N)\n',eval(char(R2)))
fprintf('R3 = %s (N)\n',eval(char(R3)))
fprintf('\n')
```

and their numerical values are:

```
R1 = -9.323149e+03 (N)
R2 = 2.474736e+04 (N)
R3 = 1.657579e+04 (N)
```

The numerical values for the axial stresses σ_1, σ_2, and σ_3 are calculated in MATLAB with:

```
sigma_1= abs(subs(sigma_1,lists,listn));
sigma_2= abs(subs(sigma_2,lists,listn));
sigma_3= abs(subs(sigma_3,lists,listn));
fprintf('sigma_1 = %s (MPa)\n',eval(char(sigma_1)))
fprintf('sigma_2 = %s (MPa)\n',eval(char(sigma_2)))
fprintf('sigma_3 = %s (MPa)\n',eval(char(sigma_3)))
fprintf('\n')
```

and the results are:

```
sigma_1 = 4.748241e+00 (MPa)
sigma_2 = 3.890039e+00 (MPa)
sigma_3 = 2.344995e+01 (MPa)
```

The displacements are calculated in MATLAB using:

```
delta1= subs(delta1,lists,listn);
delta2= subs(delta2,lists,listn);
delta3= subs(delta3,lists,listn);
```

The values are displayed in MATLAB using:

```
fprintf('delta_1 = %s (mm)\n',eval(char(delta1)))
fprintf('delta_2 = %s (mm)\n',eval(char(delta2)))
fprintf('delta_3 = %s (mm)\n',eval(char(delta3)))
fprintf('\n')
```

that is,

```
delta_1 = -2.826334e-03 (mm)
delta_2 = 1.204060e-03 (mm)
delta_3 = 1.060831e-02 (mm)
```

Example 1.4. Bars AP, PB, and PC are connected and loaded as shown in Fig. 1.22. Bar AP (bar 1) has length l_1, modulus of elasticity E_1 and a circular cross-section with radius that varies from r_{1P} at the top P to r_{1A} at the bottom A. Bar PB (bar 2) has length l_{PB}, radius $r_{PB} = r$, and modulus of elasticity $E_2 = E$. Bar PC (bar 3) has length $l_{PC} = l_{PB}$, radius $r_{PC} = r_{PB} = r$, and modulus of elasticity $E_3 = E_2 = E$. The distance between point P and base BC is l_2, and the angle of bars PB and PC with the horizontal direction is denoted by θ as shown in Fig. 1.22. Load F which is acting at point P has an upward vertical direction. Determine the load and displacement of the bars when the externally applied force F acts as shown.

Solution

For the system in Fig. 1.22, one can write the equilibrium equations. The free-body diagram shown in Fig. 1.23 contains tree unknown forces, reactions R_1, R_2, and R_3. Resolving forces horizontally, one can obtain

$$\sum F_x = 0 \Leftrightarrow R_2 \cos\theta - R_3 \cos\theta = 0. \tag{1.94}$$

Resolving forces vertically, one can obtain

$$\sum F_y = 0 \Leftrightarrow F - R_1 + R_2 \sin\theta + R_3 \sin\theta = 0. \tag{1.95}$$

The equilibrium equations are written in MATLAB with:

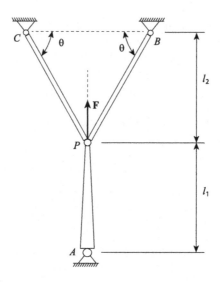

Figure 1.22 System of three connected bars under loading.

```
clear all; clc; close all
syms F R_1 R_2 R_3 l1 l2 r_1A r_1P
syms d1 d2 d3 r r3 E E_1 theta

sumFOx = R_2*cos(theta) - R_3*cos(theta);
sumFOy = F - R_1 + R_2*sin(theta) + R_3*sin(theta);
fprintf('eq1 = %s = 0\n',char(sumFOx))
fprintf('eq2 = %s =0\n',char(sumFOy))
```

Another equation is needed to determine reactions R_1, R_2, and R_3 (Fig. 1.21). Considering the deformed geometry (with exaggerated deflection) of the bars AP, PB, and PC shown in Fig. 1.24, one can write

$$\cos\alpha = \frac{\delta_3}{\delta_1} \Leftrightarrow \delta_3 = \delta_1 \cos\alpha = \delta_1 \sin\theta, \tag{1.96}$$

where $\cos\alpha = \sin\theta$ and $\delta_3 = \delta_2$ due to symmetry. The displacements δ_3 and δ_2 of bars PC and PB can be expressed using

$$\delta_2 = \frac{R_2 l_{PB}}{A_{PB} E_2} = \frac{R_2 \dfrac{l_2}{\sin\theta}}{\pi r_{PB}^2 E_2} = \frac{R_2 \dfrac{l_2}{\sin\theta}}{\pi r^2 E}, \tag{1.97}$$

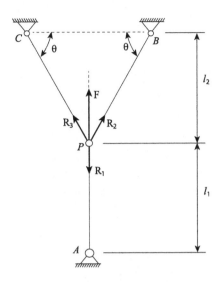

Figure 1.23 Free-body diagram of the bars under loading.

where $A_{PB} = \pi r_2^2$ is the area of the cross-section of bar PB, and $\delta_3 = \delta_2$. The displacement of bar PA can be expressed by

$$
\begin{aligned}
\delta_1 &= \int_0^{l_1} \frac{R(x)}{A(x)E_1} dx = \int_0^{l_1} \frac{R_1}{A(x)E_1} dx \\
&= \int_0^{l_1} \frac{R_1}{\pi r^2(x)E_1} dx = \frac{R_1}{\pi E_1} \int_0^{l_1} \frac{1}{r^2(x)} dx \\
&= \frac{R_1}{\pi E_1} \int_0^{l_1} \frac{1}{\left[r_{1A} - \dfrac{x}{l_1}(r_{1A} - r_{1P}) \right]^2} dx
\end{aligned}
\tag{1.98}
$$

After integrating the right-hand side of Eq. (1.98), one obtains

$$
\delta_1 = \frac{R_1 l_1}{\pi E_1 r_{1A} r_{1P}}.
\tag{1.99}
$$

From Eqs. (1.96), (1.97), and (1.99), one gets

$$
\frac{R_2 \dfrac{l_2}{\sin\theta}}{\pi r^2 E} = \frac{R_1 l_1}{\pi E_1 r_{1A} r_{1P}} \sin\theta.
\tag{1.100}
$$

Using the geometry of the deformation, the equilibrium equations are written in MATLAB with:

```
r2=r; r3=r;
A_2=pi*r2^2;
A_3=pi*r3^2;
E_2=E; E_3=E;
delta_2=R_2*12/(sin(theta)*A_2*E_2);
delta_1=R_1*l1/(pi*r_1A*r_1P*E_1);
eq3 = delta_2 - delta_1*sin(theta) ;
fprintf('eq3 = %s =0\n\n',char(eq3))
```

From Eqs. (1.94), (1.95), and (1.100), one can write the following system of three equations with three unknowns (reactions R_1, R_2, and R_3):

$$\begin{cases} R_2 \cos\theta - R_3 \cos\theta = 0, \\ F - R_1 + R_2 \sin(\theta) + R_3 \sin\theta = 0, \\ \dfrac{R_2 \dfrac{l_2}{\sin\theta}}{\pi r^2 E} = \dfrac{R_1 l_1}{\pi E_1 r_{1A} r_{1P}} \sin\theta. \end{cases} \tag{1.101}$$

Solving Eq. (1.101), one obtains

$$R_1 = \frac{F}{1 - 2\dfrac{l_1}{l_2}\dfrac{E_2}{E_1}\dfrac{r^2}{r_{1A} r_{1P}} \sin^3\theta},$$

$$R_2 = F\frac{\dfrac{l_1}{l_2}\dfrac{E_2}{E_1}\dfrac{r^2}{r_{1A} r_{1P}} \sin^2\theta}{1 - 2\dfrac{l_1}{l_2}\dfrac{E_2}{E_1}\dfrac{r^2}{r_{1A} r_{1P}} \sin^3\theta},$$

$$R_3 = F\frac{\dfrac{l_1}{l_2}\dfrac{E_3}{E_1}\dfrac{r^2}{r_{1A} r_{1P}} \sin^2\theta}{1 - 2\dfrac{l_1}{l_2}\dfrac{E_3}{E_1}\dfrac{r^2}{r_{1A} r_{1P}} \sin^3\theta}, \tag{1.102}$$

and in MATLAB:

```
sol=solve(sumFOx,sumFOy,eq3,...
    R_1,R_2,R_3);
R1new = sol.R_1;
R2new = sol.R_2;
R3new = sol.R_3;
fprintf('R1 = %s \n',char(R1new))
fprintf('R2 = %s \n',char(R2new))
fprintf('R3 = %s \n\n',char(R3new))
```

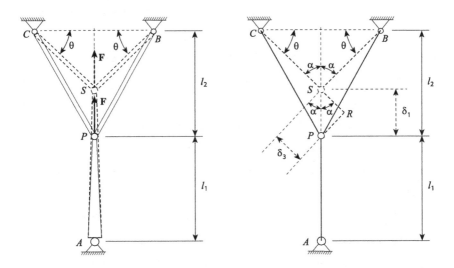

Figure 1.24 Deformed geometry (with exaggerated deflection) of the bars.

The displacements of the bars (see Fig. 1.24) are

$$\delta_3 = \delta_2 = \frac{R_1 l_1}{\pi E_1 r_{1A} r_{1P}} \sin\theta = \frac{F l_1 \sin\theta}{\pi E_1 r_{1A} r_{1P} \left(1 - 2\dfrac{l_1}{l_2}\dfrac{E_2}{E_1}\dfrac{r^2}{r_{1A} r_{1P}}\sin^3\theta\right)},$$

$$\delta_1 = \frac{R_1 l_1}{\pi E_1 r_{1A} r_{1P}} = \frac{F l_1}{\pi E_1 r_{1A} r_{1P} \left(1 - 2\dfrac{l_1}{l_2}\dfrac{E_2}{E_1}\dfrac{r^2}{r_{1A} r_{1P}}\sin^3\theta\right)}. \qquad (1.103)$$

In MATLAB, the displacement of the bars are obtained with:

```
delta_1=subs(delta_1,R_1,R1new);
delta_2=subs(delta_2,R_2,R2new);
delta_3=delta_2;
fprintf('delta_1 = %s \n',char(delta_1))
fprintf('delta_2 = %s \n',char(delta_2))
fprintf('delta_3 = %s \n\n',char(delta_3))
```

The numerical data is introduced in MATLAB using:

```
%numerical application
lists={F,r,r_1A,r_1P,l1,l2,E,E_1,theta};
listn={11500,5,7.5,10,90,120,21*10^4,...
    21*10^4,60*pi/180};
```

The numerical values for the reaction forces are calculated in MATLAB with:

```
R1 = subs(R1new,lists,listn);
R2 = subs(R2new,lists,listn);
R3 = subs(R3new,lists,listn);
fprintf('R1 = %s (N)\n',eval(char(R1)))
fprintf('R2 = %s (N)\n',eval(char(R2)))
fprintf('R3 = %s (N)\n\n',eval(char(R3)))
```

and the results are:

```
R1 = 1.703097e+04 (N)
R2 = 3.193307e+03 (N)
R3 = 3.193307e+03 (N)
```

The displacements are calculated and printed in MATLAB with:

```
delta_2=abs(subs(delta_2,lists,listn));
delta_3 = delta_2;
delta_1=abs(subs(delta_1,lists,listn));
fprintf('delta_1 = %s (mm)\n',eval(delta_1))
fprintf('delta_2 = %s (mm)\n',eval(delta_2))
fprintf('delta_3 = %s (mm)\n\n',eval(delta_3))
```

and the numerical results are:

```
delta_1 = 3.097786e-02 (mm)
delta_2 = 2.682762e-02 (mm)
delta_3 = 2.682762e-02 (mm)
```

Example 1.5. Lever 1 having length $AC = l$ is subjected to a horizontal force F applied at its end as shown in Fig. 1.25. Bar 2 denoted by AB has radius r and length d. Determine the normal and shear stresses of an element located on the free surface of road 2 at a distance h from the hexagonal base (the element is parallel to the plane determined by the axes x, y). Construct the Mohr's circle and determine the principal planes and stresses.

Solution

The normal and shear stresses of a general element (as shown in Fig. 1.1B) parallel to the plane determined by the axes x, y located at a

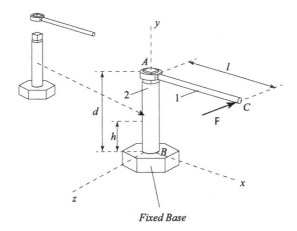

Figure 1.25 Lever under loading.

distance h from the hexagonal base can be evaluated by

$$\sigma_y = \frac{M_{h_{Ox}}r}{I_x} = \frac{4F(d-h)}{\pi r^3},$$

$$\tau_{xy} = \frac{T_{h_{Oy}}r}{J_z} = \frac{2Fl}{\pi r^3}, \tag{1.104}$$

where $I_x = \dfrac{\pi r^4}{4}$ is the second moment of area with respect to the Ox-axis, $J_z = \dfrac{\pi r^4}{2}$ is the polar moment of inertia about the axis Oz, $T_{h_{Oy}} = Fl$, and $M_{h_{Ox}} = F(d-h)$. The normal and shear stresses are calculated in MATLAB with:

```
I_x=pi*r^4/4; J_z=pi*r^4/2;
T_h0y=F*l;M_h0x=F*(d-h);
sigma_x=0; sigma_y=M_h0x*r/I_x;
tau_xy=T_h0y*r/J_z;
fprintf('The normal and shear stress are \n');
fprintf('sigma_y = %s \n',char(sigma_y))
fprintf('tau_xy = %s \n\n',char(tau_xy))
```

The element orientation and extreme values of the shear stress are calculated using Eqs. (1.7) and (1.8), and the principal stresses are calculated using Eq. (1.14). The element orientation and principal stresses are calculated in MATLAB with:

```
[sigma_max,sigma_min,radius,center_circle...
    ,phi,phi_p,sigma_theta,tau_theta]=...
    mohr2D(sigma_x,sigma_y,tau_xy,theta);

fprintf('The principal stress orientation is \n');
fprintf('tan(2*phi) = %s \n\n',char(phi_p))
fprintf('sigma_max = %s \n',char(sigma_max))
fprintf('sigma_min = %s \n',char(sigma_min))
fprintf('\n')
```

where $mohr2D(sigma_x, sigma_y, tau_{xy}, theta)$, e.g., a user created function, is:

```
function [sigma_max,sigma_min,radius,...
center_circle,phi,phi_p,sigma_theta,tau_theta]=...
    mohr2D(sigma_x,sigma_y,tau_xy,theta)
phi_p= 2*tau_xy/(sigma_x-sigma_y);
phi_radians= atan(phi_p)/2;
phi=(180*phi_radians)/(pi);
sigma_min=(sigma_x+sigma_y)/2-...
    sqrt(((sigma_x-sigma_y)/2)^2+tau_xy^2);
sigma_max=(sigma_x+sigma_y)/2+...
    sqrt(((sigma_x-sigma_y)/2)^2+tau_xy^2);
radius=sqrt(((sigma_x-sigma_y)/2)^2+tau_xy^2);
center_circle=(sigma_x+sigma_y)/2;
sigma_theta=(sigma_x+sigma_y)/2+...
    ((sigma_x-sigma_y)/2)*cos(2*theta)+...
    tau_xy*sin(2*theta);
tau_theta=-((sigma_x-sigma_y)/2)*sin(2*theta)+...
    tau_xy*cos(2*theta);
end
```

The input numerical data are introduced in MATLAB with:

```
%numerical application
lists={F,d,l,h,r};
listn={250,0.2,0.25,0.05,0.009};
```

The numerical values for the equivalent force-couple system are calculated and printed in MATLAB with:

```
T=subs(T_h0y,lists,listn);
F=subs(F,lists,listn);
```

```
M_x=subs(M_h0x,lists,listn);
fprintf('The equivalent force system is given by\n');
fprintf('T_h0y = %s (Nm)\n', char(T))
fprintf('F_h = %s (N)\n', char(F))
fprintf('M_h0x = %s (Nm)\n\n', char(M_x))
```

resulting in:

```
T_h0y = 125/2 (Nm)
F_h = 250 (N)
M_h0x = 75/2 (Nm)
```

The numerical values for the stress orientation and principal stresses are calculated and printed in MATLAB with:

```
%Mohr's circle
sigma_x=0;
sigma_y=eval(subs(sigma_y,lists,listn));
tau_xy=eval(subs(tau_xy,lists,listn));
sigma_max=eval(subs(sigma_max,lists,listn));
sigma_min=eval(subs(sigma_min,lists,listn));
radius=eval(subs(radius,lists,listn));
center_circle=eval(subs(center_circle,lists,listn));
phi=eval(subs(phi,lists,listn));
phi_p=eval(subs(phi_p,lists,listn));

fprintf('The principal stress orientation is \n');
fprintf('tan(2*phi) = %s \n', phi_p)
fprintf('or equivalently, \n')
fprintf('phi = %s degrees\n\n', phi)
fprintf('The principal stresses are \n');
fprintf('sigma_max = %f (MPa)\n', sigma_max*10^(-6))
fprintf('sigma_min = %f (MPa)\n', sigma_min*10^(-6))
```

One can obtain:

```
tan(2*phi) = -1.666667e+00
or equivalently,
phi = -2.951812e+01 degrees

The principal stresses are
sigma_max = 96.398467 (MPa)
sigma_min = -30.902605 (MPa)
```

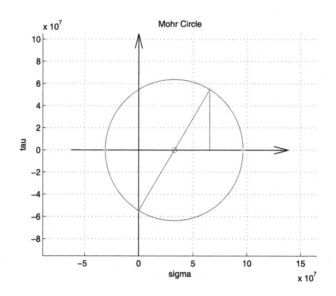

Figure 1.26 Mohr's circle.

The graphical representation of the Mohr's circle (see Fig. 1.26) is obtained in MATLAB with:

```
%Mohr's circle drawing
function [] = mohr2Ddraw(sigma_x,sigma_y,tau_xy,...
    sigma_max,sigma_min,center_circle,radius)
xlabel('sigma'), ylabel('tau')
title('Mohr's Circle')
hold on; grid on
axis equal; axis on
x_axis=quiver(center_circle-1.5*radius,0,...
    center_circle+3*radius,0,...
'Color','b','LineWidth',1.0);
set(x_axis,'Color','black');
y_axis=quiver(0,-1.5*radius,0,3.5*radius,...
'Color','b','LineWidth',1.0);
set(y_axis,'Color','black');
plot(center_circle, 0, 'o','Color', 'red');
circle(center_circle,0,radius);
text(double(radius/15),double(radius/10),'0','fontsize',12)
text(double(2*radius),double(radius/7),'sigma','fontsize',12)
text(double(radius/10),double(1.5*radius),'tau','fontsize',12)
```

```
line([sigma_y sigma_x],[tau_xy -tau_xy])
hold on
plot([sigma_x sigma_x],[-tau_xy 0],'--')
hold on
plot([sigma_y sigma_y],[tau_xy 0],'--')
plot(sigma_y,tau_xy, '+','Color', 'green');
plot(sigma_x,-tau_xy, '+','Color', 'green');
plot(sigma_max,0, '*','Color', 'red');
plot(sigma_min,0, '*','Color', 'red');
text(double(sigma_max),-double(radius/7),'sigma-max','fontsize',10)
text(double(sigma_min),-double(radius/7),'sigma-min','fontsize',10)
end
```

Example 1.6. A general stress element (as shown in Fig. 1.1B) of a structural member has $\sigma_x = 80$ MPa, $\sigma_y = 40$, and $\tau_{xy} = \tau_{yx} = 30$ MPa. Construct Mohr's circle and determine:

(a) the principal stresses,

(b) the maximum shearing stress and the associate normal stress.

Solution

The principal stresses σ_{min} and σ_{max} can be calculated from the equation $\det(A - \lambda I) = 0$ in λ, that is,

$$\begin{vmatrix} \sigma_x - \lambda & \tau_{yx} \\ \tau_{xy} & \sigma_y - \lambda \end{vmatrix} = 0, \tag{1.105}$$

where $\sigma_{min} = \lambda_1$, $\sigma_{max} = \lambda_2$, and the associated matrix A is

$$A = \begin{pmatrix} \sigma_x & \tau_{yx} \\ \tau_{xy} & \sigma_y \end{pmatrix}. \tag{1.106}$$

The principal stresses can be calculated in MATLAB with:

```
clear all; clc; close all
syms sigma_x sigma_y
syms tau_xy tau_yx

tau_xy=tau_yx;
A=[ sigma_x tau_yx; tau_xy sigma_y];
fprintf('A= '); fprintf('\n')
pretty(A)
```

The maximum shear stress is calculated using

$$\tau_{max} = \text{radius} = \frac{1}{2}(\sigma_1 - \sigma_2),\qquad(1.107)$$

where "radius" is the radius of Mohr's circle. The corresponding normal stress can be calculated as

$$\sigma = \text{center} = \frac{1}{2}(\sigma_1 + \sigma_2),\qquad(1.108)$$

where "center" is the calculated value of the center of Mohr's circle. The maximum shearing stress and the associate normal stress are calculated in MATLAB with:

```
lambda=eig(A);
sigma_min = lambda(1);
sigma_max = simplify(lambda(2));
center=1/2*(sigma_min+sigma_max);
radius=1/2*(sigma_min-sigma_max);
fprintf('The principal stresses are \n');
fprintf('sigma_min = %s \n',char(sigma_min))
fprintf('sigma_max = %s \n',char(sigma_max))
fprintf('\n')
```

The input numerical data are introduced in MATLAB with:

```
lists={sigma_x,sigma_y,tau_yx};
listn={50,-10,40};
```

The numerical values for the principal stresses are computed in MATLAB with:

```
sigma_min=subs(sigma_min,lists,listn);
sigma_max=subs(sigma_max,lists,listn);
fprintf('The principal stresses are \n');
fprintf('sigma_min = %f (MPa)\n',...
    eval(char(sigma_min)))
fprintf('sigma_max = %f (MPa)\n\n',...
    eval(char(sigma_max)))
```

resulting in:

```
sigma_min = -30.000000 (MPa)
sigma_max = 70.000000 (MPa)
```

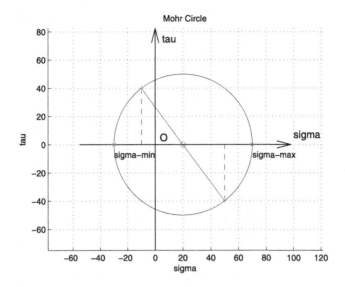

Figure 1.27 Example 1.6(b).

The maximum shearing stress and the associate normal stress are calculated and printed in MATLAB with:

```
center=subs(center,lists,listn);
radius=abs(subs(radius,lists,listn));
tau_max=subs(radius,lists,listn);
fprintf('The corresponding normal stress is \n');
fprintf('sigma = %f (MPa)\n', eval(char(center)))
fprintf('The maximum shear stress is \n');
fprintf('tau_max = %f (MPa)\n', eval(char(tau_max)))
```

and the results are:

```
The corresponding normal stress is
sigma = 20.000000 (MPa)
The maximum shear stress is
tau_max = 50.000000 (MPa)
```

The graphical representation of the Mohr's circle shown in Fig. 1.27 is obtained in MATLAB with:

```
%Mohr's circle drawing
sigma_x=subs(sigma_x,lists,listn);
```

```
sigma_y=subs(sigma_y,lists,listn);
tau_xy=subs(tau_xy,lists,listn);
mohr2Ddraw(sigma_x,sigma_y,tau_xy,sigma_max,...
    sigma_min,center,radius)
```

Example 1.7. A general stress element (as shown in Fig. 1.1A) on a structure has $\sigma_x = 30$ MPa, $\sigma_y = -30$ MPa, $\sigma_z = 70$ MPa, $\tau_{xy} = \tau_{yx} = 40$, and $\tau_{xz} = \tau_{zx} = \tau_{yz} = \tau_{zy} = 0$ MPa. Construct Mohr's circle and determine the principal stresses. Calculate the maximum shearing stress acting on the element.

Solution

The principal stresses σ_1, σ_2, and σ_3 can be calculated from the equation $\det(A - \lambda I) = 0$ in λ, that is,

$$\det(A - \lambda I) = \begin{vmatrix} \sigma_x - \lambda & \tau_{yx} & \tau_{zx} \\ \tau_{xy} & \sigma_y - \lambda & \tau_{zy} \\ \tau_{xz} & \tau_{yz} & \sigma_z - \lambda \end{vmatrix} = 0, \qquad (1.109)$$

where $\sigma_1 = \lambda_1$, $\sigma_2 = \lambda_2$, and $\sigma_3 = \lambda_3$ are the eigenvalues and

$$A = \begin{pmatrix} \sigma_x & \tau_{yx} & \tau_{zx} \\ \tau_{xy} & \sigma_y & \tau_{zy} \\ \tau_{xz} & \tau_{yz} & \sigma_z \end{pmatrix} \qquad (1.110)$$

is the eigenmatrix of Eq. (1.109). Since $\tau_{zx} = \tau_{xz} = 0$ and $\tau_{zy} = \tau_{yz} = 0$, one of the principal values of the equation is $\lambda_3 = \sigma_3$. In MATLAB the associated eigenmatrix and principal stresses are calculated and printed using:

```
clear all; clc; close all
syms sigma_x sigma_y sigma_z
syms tau_xy tau_yx tau_xz tau_zx tau_yz tau_zy

tau_xz=0; tau_zx=0;
tau_yz=0; tau_zy=0;
tau_xy=tau_yx;

[sigma1,sigma2,sigma3,center12,center23,center31,...
    radius12,radius23,radius31,tau_12,tau_23,tau_31,...
    EigValMatrix] = mohr3D(sigma_x,sigma_y,sigma_z,...
    tau_xy,tau_yx,tau_xz,tau_zx,tau_yz,tau_zy);
```

```
fprintf('The Eigen Matrix is = \n\n');
pretty(EigValMatrix);
fprintf('sigma1 = %s \n',char(sigma1))
fprintf('sigma2 = %s \n',char(sigma2))
fprintf('sigma3 = %s \n\n',char(sigma3))
```

where *mohr3D* (listed below) is a MATLAB function that calculates all the associated values related to the Mohr's circle:

```
function [sigma1,sigma2,sigma3,center12,center23,...
    center31,radius12,radius23,radius31,tau_12,...
    tau_23,tau_31,EigValMatrix] = mohr3D(sigma_x,...
    sigma_y,sigma_z,tau_xy,tau_yx,tau_xz,tau_zx,...
    tau_yz,tau_zy);
EigValMatrix= [sigma_x tau_yx tau_zx; tau_xy...
    sigma_y tau_zy; tau_xz tau_yz sigma_z];
lambda=eig(EigValMatrix);
sigma1 = lambda(1);
sigma2 = lambda(2);
sigma3 = lambda(3);
center12=1/2*(sigma1+sigma2);
center23=1/2*(sigma2+sigma3);
center31=1/2*(sigma3+sigma1);
radius12=1/2*(sigma1-sigma2);
radius23=1/2*(sigma2-sigma3);
radius31=1/2*(sigma3-sigma1);
tau_12=radius12;
tau_23=radius23;
tau_31=radius31;
end
```

One can calculate the centers of the 3 circles and their associated radii for the 3D Mohr's diagram using

$$C_{23} = \frac{1}{2}(\sigma_2 + \sigma_3), \quad C_{31} = \frac{1}{2}(\sigma_3 + \sigma_1), \quad C_{12} = \frac{1}{2}(\sigma_1 + \sigma_2), \quad (1.111)$$

and

$$R_{23} = \frac{1}{2}|\sigma_2 - \sigma_3|, \quad R_{31} = \frac{1}{2}|\sigma_3 - \sigma_1|, \quad R_{12} = \frac{1}{2}|\sigma_1 - \sigma_2|. \quad (1.112)$$

The shear and normal stresses are located outside the circle centered at C_{23}, inside the circle centered at C_{11} (largest circle) and outside the circle centered at C_{13}. The maximal shear stresses can be calculated as the largest differences between the principal values:

$$\tau_{23} = \frac{1}{2}(\sigma_2 - \sigma_3),$$

$$\tau_{13} = \frac{1}{2}(\sigma_1 - \sigma_3),$$

$$\tau_{12} = \frac{1}{2}(\sigma_1 - \sigma_2). \qquad (1.113)$$

The maximal shear stresses are calculated in MATLAB with the *mohr3D* function and printed with:

```
fprintf('tau23 = %s \n',char(radius23))
fprintf('tau31 =  %s \n',char(radius31))
fprintf('tau12 =  %s \n\n',char(radius12))
```

The largest shear stress is the maximal value of the maximal shear stresses (from 3D Mohr's circle)

$$\tau_{max} = \max(\tau_{23}, \tau_{13}, \tau_{12}). \qquad (1.114)$$

The input numerical data is introduced in MATLAB using:

```
lists={sigma_x,sigma_y,sigma_z,tau_yx};
listn={30,-30,70,40};
```

The maximal shear stresses are calculated in MATLAB using:

```
%Principal stresses
sigma_1=subs(sigma1,lists,listn);
sigma_2=subs(sigma2,lists,listn);
sigma_3=subs(sigma3,lists,listn);
```

The numerical values of the maximal shear stresses are printed in MATLAB with:

```
fprintf('sigma1 = %f (MPa)\n', eval(char(sigma_1)))
fprintf('sigma2 = %f (MPa)\n', eval(char(sigma_2)))
fprintf('sigma3 = %f (MPa)\n', eval(char(sigma_3)))
fprintf('\n')
```

that is,

```
sigma1 = 70.000000 (MPa)
sigma2 = -50.000000 (MPa)
sigma3 = 50.000000 (MPa)
```

The numerical values for the principal stresses are calculated and printed in MATLAB using:

```
tau_23=subs(tau_23,lists,listn);
tau_31=subs(tau_31,lists,listn);
tau_12=subs(tau_12,lists,listn);
fprintf('tau23 = %f (MPa)\n', eval(char(tau_23)))
fprintf('tau31 = %f (MPa)\n', eval(char(tau_31)))
fprintf('tau12 = %f (MPa)\n', eval(char(tau_12)))
fprintf('\n')
```

The MATLAB results are:

```
tau23 = -50.000000 (MPa)
tau31 = -10.000000 (MPa)
tau12 = 60.000000 (MPa)
```

The largest shear stress is calculated in MATLAB with:

```
tau_max=max([tau_23 tau_31 tau_12]);
fprintf('tau_max = %f (MPa)\n', eval(char(tau_max)))
```

resulting in:

```
tau_max = 60.000000 (MPa)
```

The graphical representation of the Mohr's circle shown in Fig. 1.28 is obtained in MATLAB using:

```
C_12=subs(center12,lists,listn);R_12=abs(tau_12);
C_23=subs(center23,lists,listn);R_23=abs(tau_23);
C_31=subs(center31,lists,listn);R_31=abs(tau_31);
%Mohr's circle drawing
mohr3Ddraw(sigma_x,sigma_y,sigma_z,...
    tau_xy,tau_yx,tau_xz,tau_zx,tau_yz,tau_zy,...
    C_12,C_23,C_31,R_12,R_23,R_31);
```

The MATLAB function *mohr3Ddraw* is listed below:

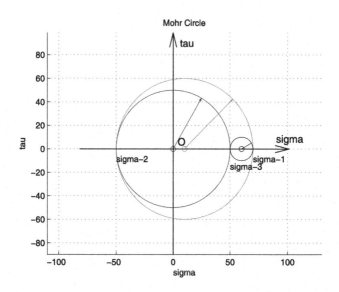

Figure 1.28 3D Mohr's circle.

```
function [] = mohr3Ddraw(sigma_x,sigma_y,sigma_z,...
    tau_xy,tau_yx,tau_xz,tau_zx,tau_yz,tau_zy,...
    center_circle1,center_circle2,center_circle3,...
    radius1,radius2,radius3);
xlabel('sigma'), ylabel('tau')
title('Mohr's Circle')
hold on; grid on
axis equal; axis on
cc=(center_circle1+center_circle2+center_circle3)/3;
rd=max([radius1 radius2 radius3]);
x_axis=quiver(cc-1.75*rd,0,cc+3*rd,0,...
'Color','b','LineWidth',1.0);
set(x_axis,'Color','black');
y_axis=quiver(0,-1.5*rd,0,3.5*rd,...
'Color','b','LineWidth',1.0);
set(y_axis,'Color','black');
plot(center_circle1, 0, 'o','Color', 'red');
text(double(radius1/15),double(radius1/10),'O','fontsize',12)
text(double(1.5*radius1),double(radius1/7),'sigma','fontsize',12)
text(double(radius1/10),double(1.5*radius1),'tau','fontsize',12)
cc1=circle(center_circle1,0,radius1);
```

```
set(cc1,'Color','red');
hold on;
plot(center_circle2, 0, 'o','Color', 'blue');
cc2=circle(center_circle2,0,radius2);
set(cc2,'Color','blue');
hold on;
plot(center_circle3, 0, 'o','Color', 'black');
cc3=circle(center_circle3,0,radius3);
set(cc3,'Color','black');
hold on;
q1=quiver(center_circle1,0,radius1*cos(pi/4),...
    radius1*sin(pi/4),0);
set(q1,'Color','red');
hold on;
q2=quiver(center_circle2,0,radius2*cos(pi/3),...
    radius2*sin(pi/3),0);
set(q2,'Color','blue');
hold on;
q3=quiver(center_circle3,0,radius3*sin(pi/3),...
    radius3*cos(pi/3),0);
set(q3,'Color','black');
text(double(center_circle1+radius1),...
    -double(radius1/7),'sigma-1','fontsize',10)
text(double(center_circle1-radius1),...
    -double(radius1/7),'sigma-2','fontsize',10)
text(double(center_circle2+radius2),...
    -double(radius1/4),'sigma-3','fontsize',10)
end
```

References

[1] E.A. Avallone, T. Baumeister, A. Sadegh, Marks' Standard Handbook for Mechanical Engineers, 11th edition, McGraw-Hill Education, New York, 2007.

[2] A. Bedford, W. Fowler, Dynamics, Addison Wesley, Menlo Park, CA, 1999.

[3] A. Bedford, W. Fowler, Statics, Addison Wesley, Menlo Park, CA, 1999.

[4] P. Boresi, O.M. Sidebottom, F.B. Selly, J.O. Smith, Advanced Mechanics of Materials, 3rd edition, John Wiley & Sons, New York, 1978.

[5] R. Budynas, K.J. Nisbett, Shigley's Mechanical Engineering Design, 9th edition, McGraw-Hill, New York, 2013.

[6] J.A. Collins, H.R. Busby, G.H. Staab, Mechanical Design of Machine Elements and Machines, 2nd edition, John Wiley & Sons, 2009.

[7] A. Ertas, J.C. Jones, The Engineering Design Process, John Wiley & Sons, New York, 1996.

[8] A.S. Hall, A.R. Holowenko, H.G. Laughlin, Schaum's Outline of Machine Design, McGraw-Hill, New York, 2013.

[9] B.G. Hamrock, B. Jacobson, S.R. Schmid, Fundamentals of Machine Elements, McGraw-Hill, New York, 1999.

[10] R.C. Juvinall, K.M. Marshek, Fundamentals of Machine Component Design, 5th edition, John Wiley & Sons, New York, 2010.

[11] K. Lingaiah, Machine Design Databook, 2nd edition, McGraw-Hill Education, New York, 2003.

[12] D.B. Marghitu, Mechanical Engineer's Handbook, Academic Press, San Diego, CA, 2001.

[13] D.B. Marghitu, M.J. Crocker, Analytical Elements of Mechanisms, Cambridge University Press, Cambridge, 2001.

[14] D.B. Marghitu, Kinematic Chains and Machine Component Design, Elsevier, Amsterdam, 2005.

[15] D.B. Marghitu, M. Dupac, H.M. Nels, Statics with MATLAB, Springer, New York, NY, 2013.

[16] D.B. Marghitu, M. Dupac, Advanced Dynamics: Analytical and Numerical Calculations with MATLAB, Springer, New York, NY, 2012.

[17] D.B. Marghitu, Mechanisms and Robots Analysis with MATLAB, Springer, New York, NY, 2009.

[18] C.R. Mischke, Prediction of stochastic endurance strength, Transaction of ASME, Journal Vibration, Acoustics, Stress, and Reliability in Design 109 (1) (1987) 113–122.

[19] R.L. Mott, Machine Elements in Mechanical Design, Prentice Hall, Upper Saddle River, NJ, 1999.

[20] W.A. Nash, Strength of Materials, Schaum's Outline Series, McGraw-Hill, New York, 1972.

[21] R.L. Norton, Machine Design, Prentice-Hall, Upper Saddle River, NJ, 1996.

[22] R.L. Norton, Design of Machinery, McGraw-Hill, New York, 1999.

[23] W.C. Orthwein, Machine Component Design, West Publishing Company, St. Paul, 1990.

[24] D. Planchard, M. Planchard, SolidWorks 2013 Tutorial with Video Instruction, SDC Publications, 2013.

[25] I. Popescu, Mechanisms, University of Craiova Press, Craiova, Romania, 1990.

[26] I.H. Shames, Engineering Mechanics – Statics and Dynamics, Prentice-Hall, Upper Saddle River, NJ, 1997.

[27] J.E. Shigley, C.R. Mischke, Mechanical Engineering Design, McGraw-Hill, New York, 1989.

[28] J.E. Shigley, C.R. Mischke, R.G. Budynas, Mechanical Engineering Design, 7th edition, McGraw-Hill, New York, 2004.

[29] J.E. Shigley, J.J. Uicker, Theory of Machines and Mechanisms, McGraw-Hill, New York, 1995.

[30] A.C. Ugural, Mechanical Design, McGraw-Hill, New York, 2004.

[31] J. Wileman, M. Choudhury, I. Green, Computation of member stiffness in bolted connections, Journal of Machine Design 193 (1991) 432–437.

[32] S. Wolfram, Mathematica, Wolfram Media/Cambridge University Press, Cambridge, 1999.

[33] Fundamentals of Engineering. Supplied-Reference Handbook, National Council of Examiners for Engineering and Surveying (NCEES), Clemson, SC, 2001.

CHAPTER TWO

Fatigue failure

Repeated, alternating, or fluctuating stress is a stress oscillating between some limits acting on a component of a system. The fatigue failure is a breakdown of the part under the action of these alternating stresses. A small crack will be sufficient to initiate fatigue failure. There are two distinct ways of a part failing due to fatigue, either a gradual growth of a crack or based on the rapid fracture of a system component.

2.1. Endurance limit

A Moore's fatigue testing machine rotates a test specimen with a constant angular speed. The test specimen is in pure bending under different weights. The fatigue strength, also called endurance strength, is the maximal reversed stress the material can withstand without failure for a considered number of cycles. For every experiment the endurance/fatigue strengths are represented as functions of the matching number of revolutions. The obtained graph known as the Wohler diagram represents the foundation of the strength–life method and depicts the nominal stress amplitude S versus the number of cycles N. The S–N diagram is usually presented on a log–log scale.

The fatigue or endurance limit S'_e of a test specimen (Fig. 2.1) for ferrous materials is the extreme value of the alternating stress that can be endlessly endured with no failure. The fatigue limit of the test specimen is represented with a horizontal line on the S–N diagram, as shown in Fig. 2.2.

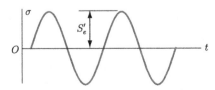

Figure 2.1 Moore's rotating beam fatigue testing system.

Machine component analysis with MATLAB®
https://doi.org/10.1016/B978-0-12-804229-8.00007-4

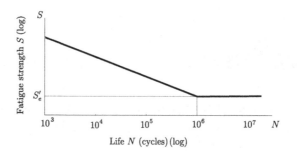

Figure 2.2 Idealized S–N diagram.

The relation between the endurance limit and the ultimate strength of the test specimen is given by [4]

$$S'_e = \begin{cases} 0.50\, S_u & \text{for } S_u \leq 200 \text{ kpsi,} \\ 100 \text{ kpsi} & \text{for } S_u > 200 \text{ kpsi,} \end{cases} \qquad (2.1)$$

or

$$S'_e = \begin{cases} 0.50\, S_u & \text{for } S_u \leq 1400 \text{ MPa,} \\ 700 \text{ MPa} & \text{for } S_u > 1400 \text{ MPa.} \end{cases} \qquad (2.2)$$

Overall, the fatigue limit of a machine component, S_e, is distinct from the one of a test specimen (sample), S'_e. The fatigue limit of a machine component, S_e, is influenced by a series of elements named modifying factors. The most important factors are the surface k_S, load k_L, gradient (size) k_G, temperature k_T, and reliability k_R. The fatigue limit of a component may be linked to the fatigue limit of the test element by the equation $S_e = k_S\, k_G\, k_L\, k_T\, k_R\, S'_e$.

Surface factor, k_S

The surface modifying factor, k_S, is related to the tensile strength, S_{ut}, of the material, and to the machine component surface finish [4] as

$$k_S = a\, S_{ut}^b. \qquad (2.3)$$

The exponent b and factor a in Eqs. (2.3) relate to the type of the surface finish (different surface conditions) [4]:

$$b = \begin{cases} -0.085 & \text{surface finish: ground,} \\ -0.256 & \text{surface finish: cold-drawn or machined,} \\ -0.718 & \text{surface finish: hot-rolled,} \\ -0.995 & \text{surface finish: forged;} \end{cases} \quad (2.4)$$

• for S_{ut} given in (kpsi)

$$a = \begin{cases} 1.34 & \text{surface finish: ground,} \\ 2.70 & \text{surface finish: cold-drawn or machined,} \\ 14.4 & \text{surface finish: hot-rolled,} \\ 39.9 & \text{surface finish: forged.} \end{cases} \quad (2.5)$$

• for S_{ut} given in (MPa)

$$a = \begin{cases} 1.58 & \text{surface finish: ground,} \\ 4.51 & \text{surface finish: cold-drawn or machined,} \\ 57.7 & \text{surface finish: hot-rolled,} \\ 272 & \text{surface finish: forged;} \end{cases} \quad (2.6)$$

Gradient (size) factor, k_G
In addition to the surface finish, the size (gradient) factor should also be considered. For bending and torsional loading, the gradient factor can be calculated as [4]

$$k_G = \begin{cases} \left(\dfrac{d}{0.3}\right)^{-0.107}, & \dfrac{11}{100} \le d \le 2 \text{ in,} \\ \dfrac{91}{100} d^{-0.157}, & 2 \le d \le 10 \text{ in,} \end{cases} \quad (2.7)$$

or

$$k_G = \begin{cases} \left(\dfrac{d}{7.62}\right)^{-0.107}, & \dfrac{279}{100} \le d \le 51 \text{ mm,} \\ \dfrac{151}{100} d^{-0.157}, & 51 \le d \le 254 \text{ mm,} \end{cases} \quad (2.8)$$

where the diameter d of the test bar was expressed both in inches and millimeters.

Eqs. (2.7) and (2.8) can be also used for a rectangular specimen or for a non-rotating round bar, in such cases an equivalent diameter d_e can be computed [4]. For a circular non-rotating shaft the equivalent diameter is

$$d_e = 0.37\, d. \tag{2.9}$$

For a rectangular section of dimensions h and b, the equivalent diameter is [4]

$$d_e = 0.808\, \sqrt{h\, b}. \tag{2.10}$$

For axial loading $k_G = 1$.

Load factor, k_L
The load modifying factor is calculated as [4]

$$k_L = \begin{cases} 0.85 & \text{for axial,} \\ 1 & \text{for bending,} \\ 0.59 & \text{for pure torsion.} \end{cases} \tag{2.11}$$

When torsion is combined with other stresses, use $k_L = 1$.

Temperature factor, k_T
The temperature factor is calculated as

$$k_T = \frac{S_{ot}}{S_{rt}}, \tag{2.12}$$

where S_{rt} and S_{ot} are the tensile strengths at room (rt) and at operating (ot) temperature, respectively.

For steel bending, torsion and axial loading, the temperature factor is [10]

$$k_T = \begin{cases} 1, & T \le 840\,°\text{F,} \\ 3.688 - 0.0032\, T, & 840\,°\text{F} < T \le 1020\,°\text{F.} \end{cases} \tag{2.13}$$

The reliability factor, k_R
One can calculate the reliability factor using

$$k_R = 1 - 0.08 z_a, \tag{2.14}$$

where

$$z_a = \begin{cases} 0, & \text{for 50\% reliability,} \\ 1.288, & \text{for 90\% reliability,} \\ 1.645, & \text{for 95\% reliability,} \\ 2.326, & \text{for 99\% reliability.} \end{cases} \tag{2.15}$$

The reliability modifying factors calculated in [10] are:

$$k_R = \begin{cases} 1, & \text{for 50\% reliability,} \\ 0.897, & \text{for 90\% reliability,} \\ 0.868, & \text{for 95\% reliability,} \\ 0.814, & \text{for 99\% reliability,} \\ 0.753, & \text{for 99.9\% reliability.} \end{cases} \qquad (2.16)$$

For the fatigue of ductile materials, the modifying factors have been reviewed in [10] and are as follows:

• for 10^3-cycle strength

$$S_f = \begin{cases} 0.9\, k_T\, S_u & \text{bending loads,} \\ 0.75\, k_T\, S_u & \text{axial loads,} \\ 0.9\, k_T\, S_{us} & \text{torsional loads,} \end{cases} \qquad (2.17)$$

where the ultimate shear strength, S_{us}, is

$$S_{us} \approx \begin{cases} 0.8\, S_u & \text{for steel,} \\ 0.7\, S_u & \text{for any other ductile material;} \end{cases} \qquad (2.18)$$

• for 10^6-cycle strength

$$S_e = k_S\, k_G\, k_L\, k_T\, k_R\, S_e'. \qquad (2.19)$$

2.2. Fluctuating stresses

The minimum and maximum stress, σ_{min} and σ_{max}, respectively, alternating stress (or stress amplitude), σ_a, stress range, σ_r, and midrange or mean stress, σ_m, are shown in Fig. 2.3. The alternating stress and mean stress are defined by

$$\sigma_a = \frac{\sigma_{max} - \sigma_{min}}{2}, \qquad (2.20)$$

$$\sigma_m = \frac{\sigma_{min} + \sigma_{max}}{2}. \qquad (2.21)$$

The stress ratios are

$$R = \frac{\sigma_{min}}{\sigma_{max}} \quad \text{and} \quad A = \frac{\sigma_a}{\sigma_m}. \qquad (2.22)$$

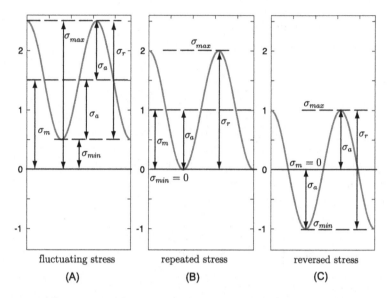

Figure 2.3 Different types of stresses: (A) sinusoidal; (B) repeated; and (C) reversed sinusoidal. From Budynas–Nisbett: Shigley's Mechanical Engineering Design, Eighth Edition, McGraw-Hill, 2006. Used with permission from McGraw Hill Inc.

In Fig. 2.3A the sinusoidal fluctuating stress is shown, while the repeated stress and the reversed sinusoidal stress are shown in Figs. 2.3B–C, respectively.

2.3. The constant life fatigue diagram

Material in this section is based on [10].

A constant life fatigue diagram is an illustration of constant life data (given yielding and various fatigue life) of mean and alternating stress as shown in Fig. 2.4. The axes (vertical and horizontal) shown in the diagram are given by $\sigma_m = 0$ and $\sigma_a = 0$, where the latter represents the static loading. The points $(0, S_e)$, $(0, S_{10^3})$, $(0, S_{10^4})$, $(0, S_{10^5})$, ... on the vertical axis are chosen directly from the S–N diagram. The ultimate tensile strength and yield strength are given by the points $(S_u, 0)$ and $(S_y, 0)$ located on the ordinate axis. The ultimate tensile strength and yield strength are given by the points $(S_u, 0)$ and $(S_y, 0)$ located on the horizontal axis. The construction lines $(0, S_e)$–$(S_u, 0)$, $(0, S_{10^3})$–$(S_u, 0)$, $(0, S_{10^4})$–$(S_u, 0)$, $(0, S_{10^5})$–$(S_u, 0)$, ... are the constant life lines (also known as Goodman

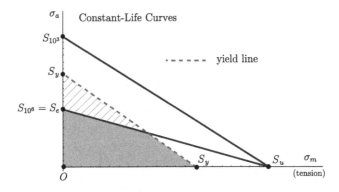

Figure 2.4 Constant life (curves) fatigue diagram.

lines) corresponding to 10^6 cycles or ∞ life, 10^3 cycles, 10^4 cycles, 10^5 cycles, and so forth.

The compressive yield strength – considered for ductile materials – is given by the point $(-S_y, 0)$. The stress at $(0, S_y)$ is bounded by $-S_y$ and $+S_y$. The points on the line $(S_y, 0)$–$(0, S_y)$ have a tensile maximum of S_y.

The gray area, in Fig. 2.4, below the lines $(0, S_e)$–$(S_u, 0)$ and $(S_y, 0)$–$(0, S_y)$ represents a life of no less than 10^6 cycles and no yielding. The hatched area above the line $(0, S_e)$–$(S_u, 0)$ and below the line $(S_y, 0)$–$(0, S_y)$ represents a life less than 10^6 cycles and no yielding. The area above the yield line, $(S_y, 0)$–$(0, S_y)$, states yielding.

In the case of torsional loads, S_{sy} and S_{us} are replacing S_y and S_u by

$$S_{sy} \approx 0.58\, S_y \quad \text{and} \quad S_{us} \approx 0.8\, S_u \quad \text{for steel.} \tag{2.23}$$

2.4. The fatigue life and arbitrarily varying forces

A component of a machine is exposed to varying load σ_i for n_i cycles, where $i = 1, 2, \ldots$ The most widely used procedure for determining the fatigue life of an element is Miner's rule (cumulative damage model) given by [10,29]:

$$\sum_{j=1}^{k} \frac{n_j}{N_i} = C \quad \text{or} \quad \frac{n_1}{N_1} + \frac{n_2}{N_2} + \cdots + \frac{n_k}{N_k} = 1, \tag{2.24}$$

where n_i, $i = 1, \ldots, k$ represent the number of cycles for considered stress amounts σ_i, $i = 1, \ldots, k$, and N_i, $i = 1, \ldots, k$ express the life (in cycles) at

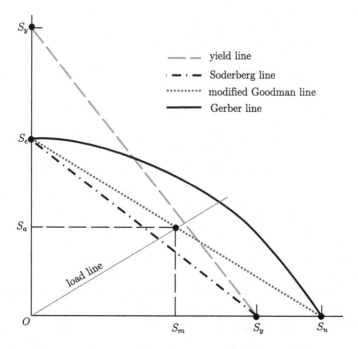

Figure 2.5 Fatigue diagram. From J.E. Shigley, and C.R. Mischke, "Mechanical Engineering Design", McGraw-Hill, Inc., 1989. Used with permission from McGraw Hill Inc.

these stresses, selected from the S–N diagram. When $c = 1$, there will be failure and experimentally $0.7 < c < 2.2$.

2.5. Variable loading failure theories

Material in this section is based on [4,29].

A fluctuating stress is characterized by a point using the mean and alternating stresses. The system will be safe if this point is located below the constant-life line. The utmost value for σ_m on the constant-life line will be the strength S_m and the utmost value for σ_a will be strength S_a, as shown in Fig. 2.5. The fatigue limit S_e is represented on the ordinate, the tensile yield strength S_y is represented on the abscissa and on the ordinate, and the (ultimate) tensile strength S_u is represented on the abscissa. The failure criteria, i.e., the yielding, Soderberg (that protects against yielding), Gerber, and the modified Goodman criteria, are depicted in the fatigue diagram shown Fig. 2.5.

The equations for the four criteria are:
- Soderberg line

$$\frac{S_m}{S_y} + \frac{S_a}{S_e} = 1.$$

(2.25)

- Modified Goodman line

$$\frac{S_m}{S_u} + \frac{S_a}{S_e} = 1.$$

(2.26)

- Yielding line

$$\frac{S_m}{S_y} + \frac{S_a}{S_y} = 1.$$

(2.27)

- Gerber parabolic relation

$$\frac{S_a}{S_e} + \left(\frac{S_m}{S_u}\right)^2 = 1.$$

(2.28)

With a safety factor, the utmost values are

$$S_a = SF\,\sigma_a \quad \text{and} \quad S_m = SF\,\sigma_m.$$

(2.29)

Soderberg equation becomes

$$\frac{\sigma_a}{S_e} + \frac{\sigma_m}{S_y} = \frac{1}{SF}.$$

(2.30)

Gerber and the modified Goodman equations become

$$\frac{SF\,\sigma_a}{S_e} + \left(\frac{SF\,\sigma_m}{S_u}\right)^2 = 1, \quad \frac{\sigma_a}{S_e} + \frac{\sigma_m}{S_u} = \frac{1}{SF}.$$

(2.31)

Some values for the tensile strength for steel (AISI = American Iron and Steel Institute) are given by MatWeb – Material Property Data, http://www.matweb.com [36]:

AISI 1006: $S_y = 285$ MPa, $S_{ut} = 330$ MPa for cold drawn,
AISI 1006: $S_y = 165$ MPa, $S_{ut} = 295$ MPa for 19–32 mm hot-rolled round bar,

AISI 1008: $S_y = 170$ MPa, $S_{ut} = 305$ MPa for 19–32 mm hot-rolled round bar,

AISI 1008: $S_y = 285$ MPa, $S_{ut} = 340$ MPa for 19–32 mm cold-drawn round bar,

AISI 1010: $S_y = 305$ MPa, $S_{ut} = 365$ MPa for cold drawn,

AISI 1010: $S_y = 180$ MPa, $S_{ut} = 325$ MPa for 19–32 mm hot-rolled round bar,

AISI 1010: $S_y = 305$ MPa, $S_{ut} = 365$ MPa for 19–32 mm cold-drawn round bar,

AISI 1012: $S_y = 310$ MPa, $S_{ut} = 370$ MPa for cold drawn,

AISI 1012: $S_y = 185$ MPa, $S_{ut} = 330$ MPa for 19–32 mm hot-rolled round bar,

AISI 1012: $S_y = 310$ MPa, $S_{ut} = 370$ MPa for 19–32 mm cold-drawn round bar,

AISI 1015: $S_y = 325$ MPa, $S_{ut} = 385$ MPa for cold drawn,

AISI 1015: $S_y = 190$ MPa, $S_{ut} = 345$ MPa for 19–32 mm hot-rolled round bar,

AISI 1015: $S_y = 325$ MPa, $S_{ut} = 425$ MPa for normalized at 925 °C (1700 °F),

AISI 1015: $S_y = 285$ MPa, $S_{ut} = 385$ MPa for annealed at 870 °C (1600 °F),

AISI 1016: $S_y = 350$ MPa, $S_{ut} = 420$ MPa for 19–32 mm cold-drawn round bar,

AISI 1016: $S_y = 205$ MPa, $S_{ut} = 380$ MPa for 19–32 mm hot-rolled round bar,

AISI 1017: $S_y = 340$ MPa, $S_{ut} = 405$ MPa for cold drawn,

AISI 1018: $S_y = 370$ MPa, $S_{ut} = 440$ MPa for cold drawn,

AISI 1018: $S_y = 275$ MPa, $S_{ut} = 475$ MPa for hot rolled, quenched,

AISI 1018: $S_y = 220$ MPa, $S_{ut} = 400$ MPa for 19–32 mm hot-rolled round bar,

AISI 1018: $S_y = 415$ MPa, $S_{ut} = 485$ MPa for 16–22 mm cold-drawn round bar,

AISI 1018: $S_y = 380$ MPa, $S_{ut} = 450$ MPa for 22–32 mm cold-drawn round bar,

AISI 1018: $S_y = 345$ MPa, $S_{ut} = 415$ MPa for 32–50 mm cold-drawn round bar,

AISI 1018: $S_y = 310$ MPa, $S_{ut} = 380$ MPa for 50–76 mm cold-drawn round bar,

AISI 1019: $S_y = 379$ MPa, $S_{ut} = 455$ MPa for 50–76 mm cold-drawn round bar,

AISI 1019: $S_y = 224$ MPa, $S_{ut} = 407$ MPa for 19–32 mm hot-rolled round bar,
AISI 1020: $S_y = 350$ MPa, $S_{ut} = 420$ MPa for cold rolled.

2.6. Examples

Example 2.1. An AISI 1015 cold-drawn steel bar has diameter $d = 20$ mm. (a) Draw the S–N curve and find the rotating bending fatigue life for 4×10^4 cycles; (b) Evaluate the constant life fatigue curves for the rotating bending load; (c) Estimate the fatigue strength corresponding to 10^3 and 4×10^4 cycles (bending and no yielding) if the part withstands repeated loads (zero to maximum).

Solution

(a) The ultimate and yield strengths are $S_u = 385$ MPa and $S_y = 325$ MPa. The endurance limit corresponding to a 10^6-cycle peak (with alternating strength) is calculated using $S_e = k_S\, k_G\, k_L\, k_T\, k_R\, S'_e$. Considering the endurance limit–ultimate strength relation given by Eq. (2.1) of a test specimen, the endurance limit for $S_u = 385$ MPa < 1400 MPa can be calculated as $S_u = 0.5\,(385) = 192.500$ MPa and $S'_e = 0.5$, or in MATLAB®:

```
Su = 385; % ultimate strength (MPa)
% Sep = 1/2 Su   (when Su<1400 MPa)
% Sep = 700 MPa (when Su>1400 MPa)
if Su < 1400    % MPa
    Sep = 1/2*Su;
else
    Sep = 700; % MPa
end
```

The surface factor is given by Eq. (2.3) and with MATLAB:

```
% surface factor kS
  a = 4.51;   % cold-drawn surface
  b = -0.256; % cold-drawn surface
kS = a * Su^b;
% kS =  0.982
```

where a and b are calculated from Eqs. (2.6), (2.5), and (2.4). The gradient (size) factor is calculated with Eq. (2.8)

$$k_G = \left(\frac{d}{7.62}\right)^{-0.107} \text{ mm for } 2.79 \le d \le 51 \text{ mm,}$$

or in MATLAB:

```
d = 20; % mm
kG = (d/7.62)^(-0.107);
% kG =  0.902
```

From Eq. (2.11) the load factor $k_L = 1$ is used for bending load. The temperature factor for bending is given by Eq. (2.13) and is $k_T = 1$. The reliability factor is calculated with Eq. (2.16) and is $k_R = 1$. Then, the endurance limit is calculated using

```
Se = kS*kG*kL*kT*kR*Sep;
% Se = 170.564 (MPa)
```

The alternating strength of 10^3-cycle peak is calculated with Eq. (2.17) as

```
% peak alternating strength 10^3 cycle strength Sf
Sf = 0.9 * kT * Su;
% Sf = 346.500 (MPa)
```

The log–log coordinates are used to plot the S–N diagram. For the abscissa $\log 10^6 = 6$, the corresponding ordinate is $\log S_e$, and for the abscissa $\log 10^3 = 3$, it results in the ordinate $\log S_f$, or in MATLAB:

```
LSe = log10(Se);
LSf = log10(Sf);
% Se = 170.564 (MPa) -> log(Se) =  2.232
% Sf = 346.500 (MPa) -> log(Sf) =  2.540
```

The S–N line in Fig. 2.6 is estimated using the line equation

$$y = mx + b,$$

with the y-intercept b and slope m computed by

$$m = \frac{\log S_e - \log S_f}{6 - 3}, \quad b = \log S_f - 3m,$$

or in MATLAB:

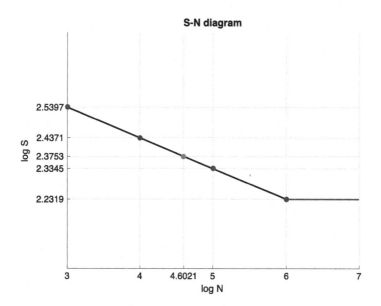

Figure 2.6 S–N diagram.

```
Lm = (LSe-LSf)/(6-3);
Lb = LSf-3*Lm;
% S-N slope: m = -0.103
% S-N y-intercept: b =  2.848
```

The log of and the 10^4-cycle peak alternating strength are

$$\log S_4 = 4\,m + b, \quad S_4 = 10^{\log S_4}.$$

The log of and the 10^5-cycle peak alternating strength are

$$\log S_5 = 5\,m + b, \quad S_5 = 10^{\log S_5}.$$

The previous calculations in MATLAB are as follows:

```
LS4 = Lb+4*Lm;
LS5 = Lb+5*Lm;
S4  = 10^LS4;
S5  = 10^LS5;
% for log(S4) =  2.437
% 10^4 cycle strength S4 = 273.589 (MPa)
% for log(S5) =  2.334
% 10^5 cycle strength S5 = 216.020 (MPa)
```

The log of and the fatigue life for $N = 4 \times 10^4$ cycles are

$$\log S_N = m \, \log(4 \times 10^4) + b, \quad S_N = 10^{\log S_N},$$

or with MATLAB:

```
NN  = 4*10^4;
LN  = log10(NN);
LSN = Lm*log10(NN)+Lb;
SN  = 10^LSN;
%  for log(SN) =  2.375
% 4*10^4 cycle strength SN = 237.314 (MPa)
```

The bending fatigue life when $N = 4 \times 10^4$ cycles is $S_N = S_{4(10^4)} = 237.314$ MPa. The results are plotted on Fig. 2.6. The graph in MATLAB is obtained with:

```
figure(1)
% S-N diagram
syms x
% eq. for Sf-Se
xn = 3:1:6;
yn = Lm*xn+Lb;
ye = Lm*x+Lb;
hold on
plot(xn,yn,'k-','LineWidth',2)
plot([6,7],[LSe,LSe],'b-','LineWidth',2)
title('S-N diagram')
ylabel('log S','FontSize',14)
xlabel('log N','FontSize',14)
axis([3 7 2 LSf+.25])
grid
ax = gca;
ax.XTick = [3,4,LN,5,6,7];
ax.YTick = [LSe,LS5,LSN,LS4,LSf];
plot(3,LSf,'bo',...
    'MarkerSize',6,'LineWidth',2,'MarkerFaceColor','b')
plot(4,LS4,'bo',...
    'MarkerSize',6,'LineWidth',2,'MarkerFaceColor','b')
plot(5,LS5,'bo',...
    'MarkerSize',6,'LineWidth',2,'MarkerFaceColor','b')
```

```
plot(LN,LSN,'ro',...
    'MarkerSize',6,'LineWidth',2,'MarkerFaceColor','r')
plot(6,LSe,'bo',...
    'MarkerSize',6,'LineWidth',2,'MarkerFaceColor','b')
```

(b) A graphical representation of the alternating/mean stress associated to the yielding and different fatigue life is represented by a fatigue diagram having as the horizontal axis $\sigma_a = 0$ (which corresponds to static loading) and as vertical axis $\sigma_m = 0$. The points S_e, S_{10^5}, $S_{4(10^4)}$, S_{10^4}, S_{10^3} on the vertical axis are chosen from the S–N diagram. The ultimate tensile and yield strengths are given by the points S_y and S_u, respectively. The points $(S_y, 0)$ and $(0, S_y)$ correspond to the tensile peak S_y. There is no yielding within the area defined by the line through the points $(S_y, 0)$ and $(0, S_y)$.

The line equations corresponding to 10^3, 10^4, $4(10^4)$, 10^5, and 10^6 cycles of life are respectively

$$\sigma_a = S_{10^3}\left(1 - \frac{\sigma_m}{S_u}\right),$$

$$\sigma_a = S_{10^4}\left(1 - \frac{\sigma_m}{S_u}\right),$$

$$\sigma_a = S_{4(10^4)}\left(1 - \frac{\sigma_m}{S_u}\right),$$

$$\sigma_a = S_{10^5}\left(1 - \frac{\sigma_m}{S_u}\right),$$

$$\sigma_a = S_e\left(1 - \frac{\sigma_m}{S_u}\right).$$

With MATLAB the equations are:

```
% constant-life fatigue curves
syms x
eq3 = Sf*(1-x/Su);
eq4 = S4*(1-x/Su);
eq5 = S5*(1-x/Su);
eq6 = Se*(1-x/Su);
eqN = SN*(1-x/Su);
eqy = Sy*(1-x/Sy);
```

and the MATLAB commands for the equation plots are:

```
figure(2)
% constant-life fatigue curves
xc = 0:1:Su;
eq3c = subs(eq3,x,xc);
eq4c = subs(eq4,x,xc);
eq5c = subs(eq5,x,xc);
eq6c = subs(eq6,x,xc);
eqNc = subs(eqN,x,xc);
eqyc = subs(eqy,x,xc);
hold on
plot(xc,eq3c,'b-','LineWidth',2)
plot(xc,eq4c,'b--','LineWidth',2)
plot(xc,eqNc,'k-','LineWidth',4)
plot(xc,eq5c,'b:','LineWidth',2)
plot(xc,eq6c,'b-.','LineWidth',2)
plot(xc,eqyc,'r-','LineWidth',2)
grid
title('constant-life fatigue diagram')
ylabel('\sigma_a (MPa)','FontSize',12)
xlabel('\sigma_m (MPa)','FontSize',12)
axis([0 Su 0 Sf])
grid on
legend('10^3','10^4','4x10^4','10^5','10^6','S_y')
ax = gca;
ax.XTick = [Sy,Su];
ax.YTick = [Se,S5,SN,S4,Sy,Sf];
set(gca, ...
'YTickLabel',num2str(get(gca,'YTick')', '%4.0f'));
set(gca, ...
'XTickLabel',num2str(get(gca,'XTick')', '%4.0f'));
```

The constant-life fatigue diagram is shown in Fig. 2.7.

(c) The case $\sigma_m = \sigma_a$ represents zero-to-maximum fluctuations of the load. The corresponding load line is the line OA shown on the constant-life fatigue diagram in Fig. 2.8. The load line OA of equation $\sigma_m = \sigma_a$ intersects the line of $N = 4(10^4)$ cycles of equation $\sigma_a = S_{4(10^4)} \left(1 - \dfrac{\sigma_m}{S_u} \right)$ in a point A

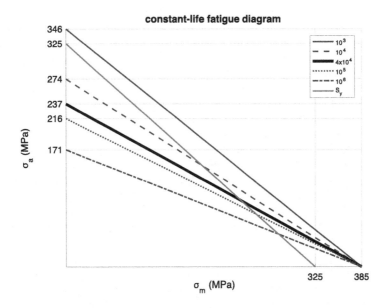

Figure 2.7 Constant-life fatigue diagram.

with coordinates $\sigma_m = \sigma_a = 146.816$ MPa. For $4(10^4)$ cycles of life, we obtain $\sigma_{max} = \sigma_a + \sigma_m = 293.633$ MPa $< S_y$.

The line OA intersect the yield line, $S_y - S_y$, of equation $\sigma_a = S_y\left(1 - \dfrac{\sigma_m}{S_y}\right)$ at the point B having coordinates $\sigma_a = \sigma_m = 162.500$ MPa. For this case $\sigma_{max} = \sigma_m + \sigma_a = 325$ MPa.

The load line OA also intersect the 10^3 cycles line of equation $\sigma_a = S_{10^3}\left(1 - \dfrac{\sigma_m}{S_u}\right)$ at the point C having coordinates $\sigma_a = \sigma_m = 182.368$ MPa. For 10^3 cycles of life, $\sigma_{max} = \sigma_a + \sigma_m = 364.737$ MPa $> S_y$. For this case $\sigma_{max} = 364.737$ MPa $\geq S_y = 325$ MPa, and this is not allowed. Since no yielding is allowed, point B is selected instead of point C, that is, $\sigma_{max} = \sigma_a + \sigma_m = 325$ MPa. The MATLAB code for the intersection of the equations is given by:

```
syms x
% load line: sigma_a = sigma_m
ms = 1;
eqL = ms*x;
% intersection of sigma_a = sigma_m with 4*10^4 cycle line
solN = solve(y-eqN,y-eqL);
```

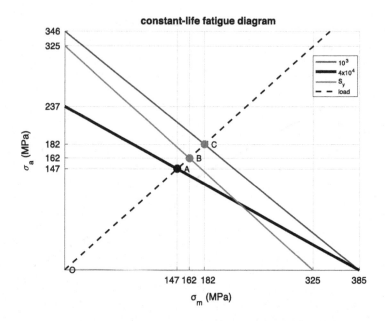

Figure 2.8 Constant-life fatigue diagram and zero-to-maximum load.

```
sigma_mN = eval(solN.x);
sigma_aN = eval(solN.y);
sigma_maxN = sigma_mN+sigma_aN;
% intersection of sigma_a = sigma_m with Sy-Sy line
soly = solve(y-eqy,y-eqL);
sigma_my = eval(soly.x);
sigma_ay = eval(soly.y);
sigma_maxy = sigma_my+sigma_ay;
% intersection of sigma_a = sigma_m with 10^3 cycle line
sol3 = solve(y-eq3,y-eqL);
sigma_m3 = eval(sol3.x);
sigma_a3 = eval(sol3.y);
sigma_max3 = sigma_m3+sigma_a3;
```

and the MATLAB graph is obtained with:

```
figure(3)
% constant-life fatigue curves
xc = 0:1:Su;
eq3c = subs(eq3,x,xc);
```

```
eqNc = subs(eqN,x,xc);
eqyc = subs(eqy,x,xc);
eqLc = subs(eqL,x,xc);

hold on
plot(xc,eq3c,'b-','LineWidth',2)
plot(xc,eqNc,'k-','LineWidth',4)
plot(xc,eqyc,'r-','LineWidth',2)
plot(xc,eqLc,'k--','LineWidth',2)
legend('10^3','4x10^4','S_y','load')
grid
title('constant-life fatigue diagram')
ylabel('\sigma_a (MPa)','FontSize',12)
xlabel('\sigma_m (MPa)','FontSize',12)
axis([0 Su 0 Sf])
grid on
ax = gca;
ax.XTick = [sigma_mN,sigma_my,sigma_m3,Sy,Su];
ax.YTick = [sigma_aN,sigma_ay,sigma_a3,SN,Sy,Sf];
set(gca, ...
'YTickLabel',num2str(get(gca,'YTick')', '%4.0f'));
set(gca, ...
'XTickLabel',num2str(get(gca,'XTick')', '%4.0f'));
plot(sigma_mN,sigma_aN,'ko',...
'MarkerSize',10,'LineWidth',2,'MarkerFaceColor','k')
plot(sigma_m3,sigma_a3,'ro',...
'MarkerSize',10,'LineWidth',2,'MarkerFaceColor','r')
plot(sigma_my,sigma_ay,'ro',...
'MarkerSize',10,'LineWidth',2,'MarkerFaceColor','r')
text(0,0,' 0','FontSize',12)
text(sigma_mN,sigma_aN,'  A','FontSize',12)
text(sigma_my,sigma_ay,'  B','FontSize',12)
text(sigma_m3,sigma_a3,'  C','FontSize',12)
```

Example 2.2. The steel shaft of the gear shown in Fig. 2.9 has $S_u = 690$ MPa and $S_y = 580$ MPa. The dimensions of the shaft and gear are: shaft diameter $d = 30$ mm, gear pitch diameter $D = 2R = 0.3$ m, and distance from the gear center O to the fillet at A is $l = OA = 0.06$ m. The shaft and fillet have a ground surface. The gear is driven at a constant speed with uniform load. The magnitudes of the gear forces at the pitch point P are $|F_x| = 600$ N, $|F_y| = 950$ N, $|F_z| = 2500$ N, and the directions are shown in Fig. 2.9. The geometric concentration factors are neglected. Evaluate the safety factor at the fillet for 90% reliability.

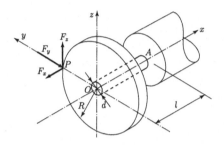

Figure 2.9 Example 2.2.

Solution

A Cartesian reference frame with the center O and axes x, y, and z is attached to the gear shown in Fig. 2.9. The gear force F acting at the point P can be expressed as

$$\mathbf{F} = F_x \mathbf{1} + F_y \mathbf{J} + F_z \mathbf{k}.$$

The position vectors $\mathbf{r}_A$ and $\mathbf{r}_P$ can be written as

$$\mathbf{r}_A = x_A \mathbf{1} + y_A \mathbf{J} + z_A \mathbf{k} = l \mathbf{1} \quad \text{and} \quad \mathbf{r}_P = x_P \mathbf{1} + y_P \mathbf{J} + z_P \mathbf{k} = R \mathbf{J}.$$

The force $\mathbf{R}$ acting at the center A of the fillet is given by

$$\mathbf{R} = -\mathbf{F} = R_x \mathbf{1} + R_y \mathbf{J} + R_z \mathbf{k},$$

where the axial force at A is $P = R_x$. A free-body diagram (forces and moments) at the fillet is shown Fig. 2.10. The moment of the force $\mathbf{F}$ with respect to the center A of the fillet is

$$\mathbf{M}_A^{\mathbf{F}} = \mathbf{r}_{AP} \times \mathbf{F} = \begin{vmatrix} \mathbf{1} & \mathbf{J} & \mathbf{k} \\ x_P - x_A & y_P - y_A & z_P - z_A \\ F_x & F_y & F_z \end{vmatrix} = \begin{vmatrix} \mathbf{1} & \mathbf{J} & \mathbf{k} \\ -l & R & 0 \\ F_x & F_y & F_z \end{vmatrix}.$$

The shaft torque with respect to longitudinal axis is $T = |M_{Ax}^{\mathbf{F}}|$, and the total bending moment due to forces on the gear is

$$M = \sqrt{\left(M_{Ay}^{\mathbf{F}}\right)^2 + \left(M_{Az}^{\mathbf{F}}\right)^2}.$$

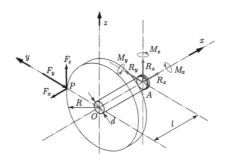

Figure 2.10 Free-body diagram of Example 2.2.

The MATLAB commands for the forces and moments are as follows:

```
Fx = -600; % (N)
Fy = -950; % (N)
Fz = 2500; % (N)
d = 0.03;    % (m)
R = 0.30/2; % (m)
l = 0.06;    % (m)
rA_ = [l, 0, 0];
rP_ = [0, R, 0];
F_  = [Fx, Fy, Fz];
RA_ = F_;   Rx = RA_(1); Ry = RA_(2); Rz = RA_(3);
P=Rx;
MA_ = cross(rP_-rA_, F_);
Rx = RA_(1); Ry = RA_(2); Rz = RA_(3);
Mx = MA_(1); T = Mx;
My = MA_(2);
Mz = MA_(3);
M = sqrt(My^2+Mz^2);
```

and the results are:

```
axial load P = Rx = -600.000 (N)
Ry = -950.000 (N)
Rz = 2500.000 (N)
torque Mx = T = 375.000 (N m)
My = 150.000 (N m)
Mz = 147.000 (N m)
total bending moment M = 210.021 (N m)
```

The stress components at the fillet are

$$\tau = \frac{16\,T}{\pi\,d^3}, \quad \sigma_b = \frac{32\,M}{\pi\,d^3}, \quad \text{and} \quad \sigma_a = \frac{4\,P}{\pi\,d^2}.$$

Unvarying operating conditions (steady-state) consider the bending stress completely reversed and the axial and torsional stresses constant. Bending provides an alternating stress. The equivalent mean bending stress, σ_{em}, is given by [10]

$$\sigma_{em} = \frac{\sigma_m}{2} + \sqrt{\tau_m^2 + \left(\frac{\sigma_m}{2}\right)^2} = \frac{\sigma_a}{2} + \sqrt{\tau^2 + \left(\frac{\sigma_a}{2}\right)^2}, \qquad (2.32)$$

and the equivalent alternating bending stress is

$$\sigma_{ea} = \sigma_b. \qquad (2.33)$$

The MATLAB commands for the mean and alternating stresses are

```
tau = 16*T/(pi*d^3);
tau = tau*10^-6; % mega (M)
% tau = 70.736 (MPa)


sigmab = 32*M/(pi*d^3);
sigma_b = sigmab*10^-6;
% sigma_b = 79.232 (MPa)


sigmax = abs(P)/(pi*d^2/4);
sigma_a = sigmax*10^-6;
% sigma_a =  0.849 (MPa)


% mean stress
sigmam = sigma_a;
sigma_em = sigmam/2+sqrt(tau^2+(sigmam/2)^2);
% sigma_em = 71.161 (MPa)


% alternating stress
sigmaa = sigma_b;
sigma_ea = sigmaa;
% sigma_ea = 79.232 (MPa)
```

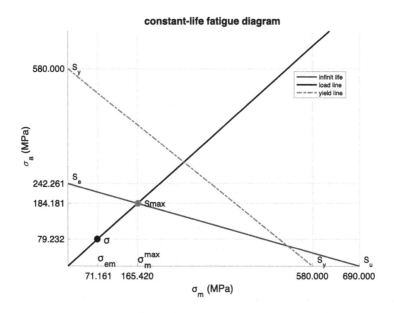

Figure 2.11 Constant life fatigue diagram of Example 2.2.

A constant life fatigue diagram of the test specimen having the endurance limit $S'_e = 0.5\,S_u$, $S_u < 1400$ MPa is shown in Fig. 2.11. The surface factor, k_S, is calculated for ground surface. The gradient (size) factor, k_G, is calculated for $d = 30$ mm where 2.79 mm $< d < 51$ mm. The reliability factor for 90% reliability is $k_R = 0.897$, the temperature and load factors, k_T and k_L, are both equal to 1. The endurance limit given by $S_e = k_S\ k_G\ k_L\ k_R\ k_T\ S'_e$ is expressed in MATLAB with:

```
if Su < 1400    % MPa
    Sep = 0.5*Su; else Sep = 700; % MPa
end
% Sep = 345.000 (MPa)
% surface factor kS
 a = 1.58;  b = -0.085; % ground
kS = a*Su^b;
% kS =  0.906
d = 30; % (mm)
kG = (d/7.62)^(-0.107);
% kG =  0.864
kL = 1; %  load factor
```

```
kR = 0.897; % 90% reliability
kT = 1; % temperature factor
Se=kS*kG*kL*kR*kT*Sep; % endurance limit
% Se = 242.261 (MPa)
```

The Goodman line equation of infinite-life (10^6) (in the bending case) is

$$\sigma_a = m\,\sigma_m + S_e,$$

where $m = (Se - 0)/(0 - Su)$. The "operating point" represented on the diagram shown in Fig. 2.11 relates to the equivalent and alternating bending stresses (σ_{em}, σ_{ea}). The load line drawn through the "operating point" (σ_{em}, σ_{ea}) and the origin O has the equation

$$\sigma_a = \frac{\sigma_{ea}}{\sigma_{em}}\,\sigma_m.$$

The intersection of the Goodman line for infinite-life and the load line gives the "failure point". The equivalent mean bending stress for the "failure point" is σ_m^{max}. The safety factor is calculated with

$$SF = \frac{\sigma_m^{max}}{\sigma_{em}}.$$

The MATLAB commands for the safety factor are:

```
% line slope for 10^6 cycles
m=(Se-0)/(0-Su);
% m = -0.351
syms x
% 10^6 cycle line
y   = m*x+Se;
% load line
y1 = (sigma_ea/sigma_em)*x;
% yield line
y2 = - x + Sy;

% intersection of load line with 10^6 cycle line
sigmacomp = solve(y1 == y, x);
sigmammax = eval(sigmacomp);
sigmaamax = m*sigmammax + Se;
```

```
% sigma_m max = 165.420 (MPa)
% sigma_a max = 184.181 (MPa)

SF = sigmammax/sigma_em;
% SF =  2.325
```

The MATLAB commands for plotting the graph are:

```
% plot the graph
x = 0:Su/10:Su;
y = m*x + Se;
y1 = (sigma_ea/sigma_em)*x;
y2 = - x + Sy;

figure(1)
hold on
plot(x,y,'b-','LineWidth',2)
plot(x,y1,'k-','LineWidth',2)
plot(x,y2,'r-.','LineWidth',2)
title('constant-life fatigue diagram')
ylabel('\sigma_a (MPa)','FontSize',12)
xlabel('\sigma_m (MPa)','FontSize',12)
axis([0 Su 0 Su])
legend('10^6','load','S_y')
ax = gca;
ax.XTick = [sigma_em,sigmammax,Sy,Su];
ax.YTick = [sigma_ea,sigmaamax,Se,Sy];
set(gca, ...
'YTickLabel',num2str(get(gca,'YTick')', '%4.3f'));
set(gca, ...
'XTickLabel',num2str(get(gca,'XTick')', '%4.3f'));
plot(sigma_em,sigma_ea,'ko',...
'MarkerSize',8,'LineWidth',2,'MarkerFaceColor','k')
plot(sigmammax,sigmaamax,'ro',...
'MarkerSize',8,'LineWidth',2,'MarkerFaceColor','r')
grid
text(Sy,15,'  S_y','FontSize',12)
text(0,Sy,'  S_y','FontSize',12)
text(Su,15,'  S_u','FontSize',12)
text(0,Se+15,'  S_e','FontSize',12)
```

```
text(sigmammax, sigmaamax, '   Smax','FontSize',12)
text(sigma_em, sigma_ea, '  \sigma','FontSize',16)

text(sigma_em, 20,'\sigma_{em}','FontSize',16)
text(sigmammax, 20,'\sigma_{m}^{max}','FontSize',16)
```

Example 2.3. The ultimate tensile and yield strengths for a steel bar with the diameter 2 in are $S_u = 150$ kpsi and $S_y = 120$ kpsi, respectively. The bar which is subjected to torsion has a ground surface finish. The fluctuating stress for 10 s of operation includes a total of 5 cycles. The first two cycles have a minimum stress equal to $\tau_{minI} = 8$ kpsi and maximum stress equal to $\tau_{maxI} = 60$ kpsi, and the next three cycles have the minimum stress equal to $\tau_{minII} = 12$ kpsi and maximum stress equal to $\tau_{maxII} = 68$ kpsi. Determine the fatigue life of the bar which is assumed to operate continuously.

Solution

The alternating and mean stresses for the first two cycles of fluctuation ($n_I = 2$) are calculated using

$$\tau_{mI} = \frac{\tau_{maxI} + \tau_{minI}}{2} \quad \text{and} \quad \tau_{aI} = \frac{\tau_{maxI} - \tau_{minI}}{2},$$

where τ_{minI} is the minimum stress and τ_{maxI} is the maximum stress.

The alternating and mean stresses for the first two cycles of fluctuation ($n_{II} = 3$) are calculated using

$$\tau_{mII} = \frac{\tau_{maxII} + \tau_{minII}}{2} \quad \text{and} \quad \tau_{aII} = \frac{\tau_{maxII} - \tau_{minII}}{2},$$

where τ_{minII} is the minimum stress and τ_{maxII} is the maximum stress.

The point with coordinates (τ_{mI}, τ_{aI}) on the τ_m–τ_a diagram and the point $\tau_m = S_{us} = 0.8 \, S_u$ located on the horizontal axis are connected by a straight line. The slope of the line equation is calculated using

$$m_I = \frac{\tau_{aI}}{\tau_{mI} - S_{us}}.$$

Setting $\tau_m = 0$ gives the y-intercept (intersection of the line and vertical axis) as

$$S_I = -m_I \, S_u.$$

For the three cycles of fluctuation ($n_{II} = 3$), the point with coordinates (τ_{mII}, τ_{aII}) on the $\tau_m-\tau_a$ plot and the point $\tau_m = S_{us}$ located on the horizontal axis are connected by a straight line. The slope of this second line is calculated using

$$m_{II} = \frac{\tau_{aII}}{\tau_{mII} - S_u},$$

and the intersection with the vertical axis ($\tau_m = 0$) is the y-intercept

$$S_{II} = -m_{II} S_{us}.$$

The lines S_I-S_{us} and $S_{II}-S_{us}$ are Goodman lines (constant life) and the corresponding fatigue lives are the points S_I and S_{II} as shown in Fig. 2.12. The MATLAB program to calculate S_I and S_{II} is:

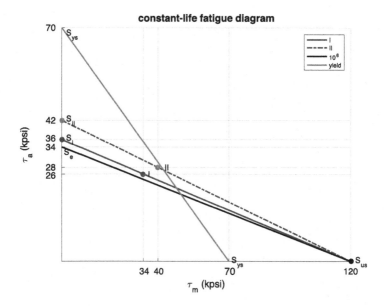

Figure 2.12 Constant life fatigue diagram of Example 2.3.

```
Su = 150; % ultimate strength (kpsi)
Sy = 120; % yield strength (kpsi)
d = 1; % (in)

t = 10; % (s) operation time
nI  = 2; % cycles
nII = 3; % cycles
```

```
tau_maxI = 60; % (kpsi)
tau_minI =  8; % (kpsi)

tau_maxII = 68; % (kpsi)
tau_minII = 12 ; % (kpsi)

tau_aI = (tau_maxI-tau_minI)/2;
tau_mI = (tau_maxI+tau_minI)/2;
% tau_aI = 26.000 (kpsi)
% tau_mI = 34.000 (kpsi)

tau_aII = (tau_maxII-tau_minII)/2;
tau_mII = (tau_maxII+tau_minII)/2;
% tau_aII = 28.000 (kpsi)
% tau_mII = 40.000 (kpsi)

Sus = 0.8*Su;
Sys = 0.58*Sy;
% Sus = 120.000 (kpsi)
% Sys = 69.600 (kpsi)

% I constant-life fatigue line slope
mI = tau_aI/(tau_mI-Sus);
% mI = -0.302
% I constant-life fatigue line y-intercept
SI = -mI*Sus;
% SI = 36.279 (kpsi)

% II constant-life fatigue line slope
mII = tau_aII/(tau_mII-Sus);
% mII = -0.350
% II constant-life fatigue line y-intercept
SII = -mII*Sus;
% SII = 42.000 (kpsi)
```

The number of cycles corresponding to S_I and S_{II} fatigue lives are resolved from the S–N diagram (in log–log coordinates). The MATLAB program for the S–N diagram is:

```
% 10^6 cycle strength (endurance limit) Se
% test specimen Se' endurance limit
% Se'=0.5 Su (for Su<200 kpsi); Se'=100 kpsi (for Su>200 kpsi)
% bending, axial, torsion: Se'=0.5 Su
if Su < 200    % kpsi
   Sep = 0.5*Su;
else
   Sep = 100; % kpsi
end
% Sep = 75.000 (kpsi)


%  endurance limit modifying factors
% (bending, axial, torsion) surface factor kS
 a = 1.34;   % ground surface
 b = -0.085; % ground surface
kS = a * Su^b;
% kS =  0.875

% size (gradient) factor kG
kG = (d/0.3)^(-0.107);
% kG =  0.879

% load factor kL
kL=0.59; % torsion

kT = 1; % temperature factor
kR = 1; % reliability factor

% Se = kS kG kL kT kR Sep  - endurance limit
Se = kS*kG*kL*kT*kR*Sep;
% Se = 34.049 (kpsi)

% peak alternating strength
% 10^3 cycle strength Sf
% bending load:  Sf = 0.9 kT Su;
% axial load:  Sf = 0.75 kT Su;
% torsional load: Sf = 0.9 kT Sus;
% Sus = 0.8 Su (for steel); Sus = 0.7 Su (other)
Sf = 0.9*kT*Sus;
% Sf = 108.000 (kpsi)
```

In the S–N diagram, for $\log 10^3 = 3$ we have $\log S_f$, and for $\log 10^6 = 6$ we have $\log S_e$. The S–N line equation (using log–log coordinates) is $y = mx + b$, and the numerical values are obtained with MATLAB using:

```
% S-N diagram
% eq. for Sf-Se
% y = Lm*x + Lb
LSe = log10(Se);
% log(Se) =  1.532
LSf = log10(Sf);
% log(Sf) =  2.033
Lm = (LSe-LSf)/(6-3);
% S-N line slope: m = -0.167
Lb = LSf-3*Lm;
% S-N line y-intercept: b =  2.535
```

The S–N diagram is plotted in Fig. 2.13. The numbers of cycles, N_I and N_{II}, corresponding to S_I and S_{II} are calculated from the S–N diagram using the MATLAB commands:

```
LSI = log10(SI);
% log(SI) =  1.560
LNI = (LSI-Lb)/Lm;
% log(NI) =  5.835
NI  = 10^LNI;
% NI = 6.841e+05 cycles
LSII = log10(SII);
% log(SII) =  1.623
LNII = (LSII-Lb)/Lm;
% log(NII) =  5.455
NII  = 10^LNII;
% NII = 2.848e+05 cycles
```

The results are shown in Fig. 2.13. The total life due to the cycles I and II can be calculated with

$$\frac{n_I\, p}{N_I} + \frac{n_{II}\, p}{N_{II}} = 1,$$

where p is the life in periods of $t = 10$ s duration and the estimated total life corresponds to $Li = pt$. The life of the part is obtained by

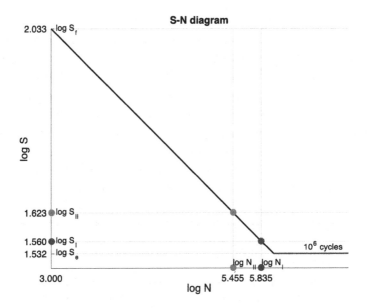

Figure 2.13 Example 2.3. S–N diagram.

```
% Miner's rule
% p nI/NI + p nII/NII = 1;
% p life in periods of t seconds
p = 1/(nI/NI + nII/NII);
Li = t*p;
% life of part =743117.131(s)=12385.286(min)=206.421(h)
```

References

[1] E.A. Avallone, T. Baumeister, A. Sadegh, Marks' Standard Handbook for Mechanical Engineers, 11th edition, McGraw-Hill Education, New York, 2007.

[2] A. Bedford, W. Fowler, Dynamics, Addison Wesley, Menlo Park, CA, 1999.

[3] A. Bedford, W. Fowler, Statics, Addison Wesley, Menlo Park, CA, 1999.

[4] R. Budynas, K.J. Nisbett, Shigley's Mechanical Engineering Design, 9th edition, McGraw-Hill, New York, 2013.

[5] J.A. Collins, H.R. Busby, G.H. Staab, Mechanical Design of Machine Elements and Machines, 2nd edition, John Wiley & Sons, 2009.

[6] A. Ertas, J.C. Jones, The Engineering Design Process, John Wiley & Sons, New York, 1996.

[7] A.S. Hall, A.R. Holowenko, H.G. Laughlin, Schaum's Outline of Machine Design, McGraw-Hill, New York, 2013.

[8] B.G. Hamrock, B. Jacobson, S.R. Schmid, Fundamentals of Machine Elements, McGraw-Hill, New York, 1999.

[9] R.C. Hibbeler, Engineering Mechanics, Prentice-Hall, Upper Saddle River, NJ, 2010.

[10] R.C. Juvinall, K.M. Marshek, Fundamentals of Machine Component Design, 5th edition, John Wiley & Sons, New York, 2010.

[11] K. Lingaiah, Machine Design Databook, 2nd edition, McGraw-Hill Education, New York, 2003.

[12] D.B. Marghitu, Mechanical Engineer's Handbook, Academic Press, San Diego, CA, 2001.

[13] D.B. Marghitu, M.J. Crocker, Analytical Elements of Mechanisms, Cambridge University Press, Cambridge, 2001.

[14] D.B. Marghitu, Kinematic Chains and Machine Component Design, Elsevier, Amsterdam, 2005.

[15] D.B. Marghitu, M. Dupac, N.H. Madsen, Statics with MATLAB, Springer, New York, NY, 2013.

[16] D.B. Marghitu, M. Dupac, Advanced Dynamics: Analytical and Numerical Calculations with MATLAB, Springer, New York, NY, 2012.

[17] D.B. Marghitu, Mechanisms and Robots Analysis with MATLAB, Springer, New York, NY, 2009.

[18] C.R. Mischke, Prediction of stochastic endurance strength, Transaction of ASME, Journal Vibration, Acoustics, Stress, and Reliability in Design 109 (1) (1987) 113–122.

[19] J.L. Meriam, L.G. Kraige, Engineering Mechanics: Dynamics, John Wiley & Sons, New York, 2007.

[20] R.L. Mott, Machine Elements in Mechanical Design, Prentice-Hall, Upper Saddle River, NJ, 1999.

[21] W.A. Nash, Strength of Materials, Schaum's Outline Series, McGraw-Hill, New York, 1972.

[22] R.L. Norton, Machine Design, Prentice-Hall, Upper Saddle River, NJ, 1996.

[23] R.L. Norton, Design of Machinery, McGraw-Hill, New York, 1999.

[24] W.C. Orthwein, Machine Component Design, West Publishing Company, St. Paul, 1990.

[25] D. Planchard, M. Planchard, SolidWorks 2013 Tutorial with Video Instruction, SDC Publications, 2013.

[26] C.A. Rubin, The Student Edition of Working Model, Addison-Wesley Publishing Company, Reading, MA, 1995.

[27] A.S. Seireg, S. Dandage, Empirical design procedure for the thermodynamic behavior of journal bearings, Journal of Lubrication Technology 104 (1982) 135–148.

[28] I.H. Shames, Engineering Mechanics – Statics and Dynamics, Prentice-Hall, Upper Saddle River, NJ, 1997.

[29] J.E. Shigley, C.R. Mischke, Mechanical Engineering Design, McGraw-Hill, New York, 1989.

[30] J.E. Shigley, C.R. Mischke, R.G. Budynas, Mechanical Engineering Design, 7th edition, McGraw-Hill, New York, 2004.

[31] J.E. Shigley, J.J. Uicker, Theory of Machines and Mechanisms, McGraw-Hill, New York, 1995.

[32] A.C. Ugural, Mechanical Design, McGraw-Hill, New York, 2004.

[33] J. Wileman, M. Choudhury, I. Green, Computation of member stiffness in bolted connections, Journal of Machine Design 193 (1991) 432–437.

[34] S. Wolfram, Mathematica, Wolfram Media/Cambridge University Press, Cambridge, 1999.
[35] Fundamentals of Engineering. Supplied-Reference Handbook, National Council of Examiners for Engineering and Surveying (NCEES), Clemson, SC, 2001.
[36] MatWeb – material property data, http://www.matweb.com/.

CHAPTER THREE

Screws

3.1. Introduction

Fasteners can be threaded, fixed, and locking, while threaded fasteners involve different screws. Fig. 3.1A shows an external helical thread coiled on the surface of a cylinder. The thread of an external screw, shown in Fig. 3.1, is described by: the pitch, p, which is defined by the length,

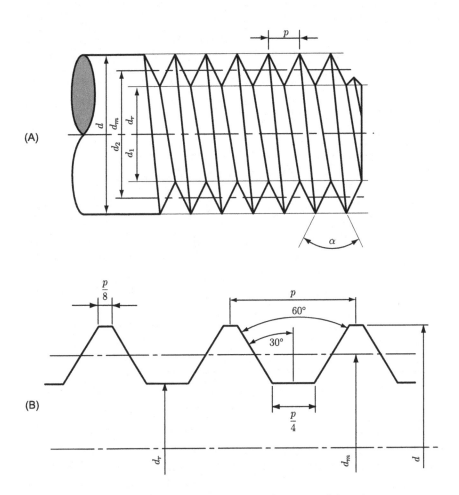

Figure 3.1 Helical thread and standard geometry of an external thread.

Machine component analysis with MATLAB®
https://doi.org/10.1016/B978-0-12-804229-8.00008-6

parallel to the axis of the screw, between the crest of one thread to the adjacent crest; the major diameter, d, which is the outside diameter; the minor diameter, d_r or d_1, which is the lower diameter; the pitch diameter, d_m or d_2, that is, the fictitious diameter for which the cuts and flanks of the threads are identical; and the lead, l, which is the axial distance the screw advances during a single 360° rotation. A single start screw or a single-threaded screw has the lead equal to the pitch. A screw with multiple threads has multiple edges next to each other: the double-start or double-threaded screw has the lead $l = 2p$ (for a complete rotation the screw moves 2 times the pitch). A right-handed screw when rotated clockwise moves away from the observer. The threads are in general right- and left-handed and must be identified. Fig. 3.2A represents a single thread right-handed screw, and a double-threaded left-handed screw is depicted in Fig. 3.2B. Fig. 3.2A shows also the lead angle, λ, which is defined as the angle of the helix of the screw and the rotation plane by $\lambda = \arctan \dfrac{l}{\pi d_m}$. The helix angle is calculated as $(\pi/2 - \lambda)$ or $\arctan \dfrac{\pi d_m}{l}$.

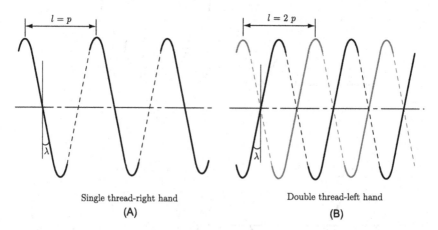

Single thread-right hand

(A)

Double thread-left hand

(B)

Figure 3.2 Single thread right-handed screw, lead angle λ, and a double-thread left-handed screw.

Metric or ISO threads are defined by M followed by the major diameter (mm) and pitch (mm/thread). Unified system threads are specified by the major diameter (in), number of treads per inch, and thread series: UNC Unified National Coarse, UNEF Unified National Extra Fine, UNF Unified National Fine, UNS Unified National Special, UNR Unified National Round (round root).

3.2. Power transmission screws

The power transmission screws transform rotational into linear motion. They are used in screw jacks, tensile devices, presses, and precise machines. Fig. 3.3 shows the Acme and square threads that are employed as power transmission screws.

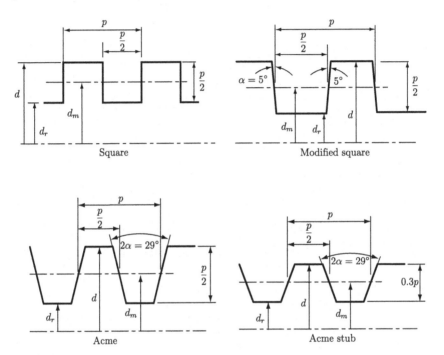

Figure 3.3 Square and Acme threads.

Fig. 3.4 shows the helix of a single square-threaded power screw with the pitch diameter d_m, pitch p, and lead angle λ. For one turn, πd_m, the lead of the thread, l, is the height of a right triangle and the "ridge" of the thread is the hypotenuse.

A compressive force, F, which is parallel to the axis of the screw, acts on the power screw as shown in Fig. 3.4. The following equations are applied for square threads.

The force for lifting the load F is P_r, and it is considered positive. The friction force acts against the motion of the screw and has the value $F_f = \mu N$, where μ is the coefficient friction and N is the normal force. For raising the load, the equilibrium equations are

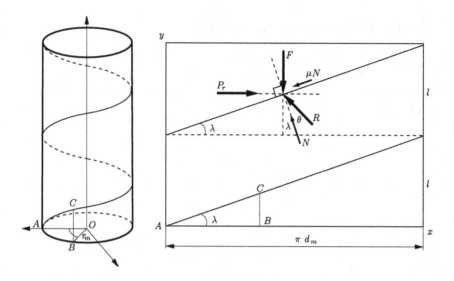

Figure 3.4 Helix of a single square-threaded screw.

$$\sum F_x = P_r - N\sin\lambda - \mu N\cos\lambda = 0,$$
$$\sum F_y = F + \mu N\sin\lambda - N\cos\lambda = 0, \qquad (3.1)$$

and the force to raise the load is

$$P_r = \frac{F(\sin\lambda + \mu\cos\lambda)}{\cos\lambda - \mu\sin\lambda}. \qquad (3.2)$$

Symbolically the force to raise the load is obtained from the MATLAB®
program:

```
syms F Pr Pl N mu lambda
% to raise the load
Fx = Pr-N*sin(lambda)-mu*N*cos(lambda);
Fy = F+mu*N*sin(lambda)-N*cos(lambda);
solPr = solve(Fx,Fy,Pr,N);
Prs = solPr.Pr;
pretty(Prs)
% F (sin(lambda) + mu cos(lambda))
% -------------------------------
%    cos(lambda) - mu sin(lambda)
```

To lower the load, the force, P_l, with $P_l < 0$, is determined from

$$\sum F_x = -P_l - N \sin\lambda + \mu N \cos\lambda = 0,$$

$$\sum F_y = F - \mu N \sin\lambda - N \cos\lambda = 0. \tag{3.3}$$

The force necessary to lower the load is

$$P_l = \frac{F(\mu\cos\lambda - \sin\lambda)}{\cos\lambda + \mu\sin\lambda}, \tag{3.4}$$

or symbolically with MATLAB:

```
% to lower the load
Fx = -P1-N*sin(lambda)+mu*N*cos(lambda);
Fy = F-mu*N*sin(lambda)-N*cos(lambda);
solP1 = solve(Fx,Fy,P1,N);
P1s = solP1.P1;
pretty(P1s)
%    F (sin(lambda) - mu cos(lambda))
% - -------------------------------
%      cos(lambda) + mu sin(lambda)
```

From the definition of the lead angle, $\tan\lambda = l/(\pi d_m)$, so a necessary moment to raise the load is

$$M_r = P_r \frac{d_m}{2} = \frac{Fd_m}{2}\left(\frac{l + \pi\mu d_m}{\pi d_m - \mu l}\right), \tag{3.5}$$

and a necessary moment to lower the load is

$$M_l = \frac{Fd_m}{2}\left(\frac{\pi\mu d_m - l}{\pi d_m + \mu l}\right). \tag{3.6}$$

The screw is *self-locking* when $M_l > 0$, and the constraint for self-locking is

$$\pi\mu d_m > l, \tag{3.7}$$

or

$$\mu > \tan\lambda. \tag{3.8}$$

If the friction is small and the lead is large, the external load can lower itself without any lowering moment. For the frictionless case ($\mu = 0$), the moment to raise the load is

$$M_0 = \frac{Fl}{2\pi}. \tag{3.9}$$

The screw efficiency e is obtained as

$$e = \frac{M_0}{M_r} = \frac{Fl}{2\pi M_r}. \tag{3.10}$$

Next a similar expression for a square-threaded jack is obtained. The axial load F and the moment M about the screw axis act on the screw with the mean radius r_m. The force of the thread on the screw thread is R as shown in Fig. 3.4. The angle of friction, θ, is between the normal to the thread, N, and the reaction force, R, and given by

$$\tan\theta = \mu = \frac{F_f}{N}.$$

The axial load F is calculated from the force equilibrium as

$$F = R\cos(\lambda + \theta),$$

where $\tan\lambda = l/(2\pi r_m)$. The moment of the reaction force R with respect to the vertical axis is $M = Rr_m \sin(\lambda + \theta)$. The moment required to raise the load is

$$M = M_r = Fr_m \tan(\lambda + \theta). \tag{3.11}$$

In a similar way the moment required to lower the load is

$$M = M_l = Fr_m \tan(\theta - \lambda). \tag{3.12}$$

The screw self-locking condition is $\theta > \lambda$.

The Acme thread has a normal thread angle, α_n, measured in the normal plane and calculated from

$$\tan\alpha_n = \tan\alpha \cos\lambda, \tag{3.13}$$

where α is the thread angle measured in the axial plane and λ is the lead angle. For the Acme threads, the moment required to raise the load is

$$M_r = P_r \frac{d_m}{2} = \frac{F d_m}{2} \left(\frac{\mu + \tan \lambda \cos \alpha_n}{\cos \alpha_n - \mu \tan \lambda} \right), \qquad (3.14)$$

or

$$M_r = \frac{F d_m}{2} \left(\frac{\mu \pi d_m + l \cos \alpha_n}{\pi d_m \cos \alpha_n - \mu l} \right), \qquad (3.15)$$

and the moment required to lower the load is

$$M_l = \frac{F d_m}{2} \left(\frac{\mu \pi d_m - l \cos \alpha_n}{\pi d_m \cos \alpha_n + \mu l} \right). \qquad (3.16)$$

The screws with square threads are more difficult to machine than the Acme threads but they are more efficient than the Acme threads.

A thrust collar can be attached to a power screw as shown in Fig. 3.5, and then a friction moment is added

$$M_c = \frac{F \mu_c d_c}{2}, \qquad (3.17)$$

where μ_c is the coefficient of collar friction and d_c is the mean collar diameter. The preferred pitches for Acme threads are [29]:

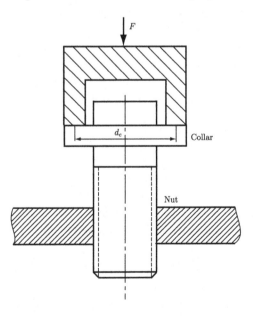

Figure 3.5 Thrust collar.

$$p = \begin{cases} 1/16 \text{ in} & \text{for } d = 0.25 \text{ in}, \\ 1/10 \text{ in} & \text{for } d = 0.50 \text{ in}, \\ 1/6 \text{ in} & \text{for } d = 0.75 \text{ in}, \\ 1/5 \text{ in} & \text{for } d = 1 \text{ in}, \ d = 1.25 \text{ in}, \\ 1/4 \text{ in} & \text{for } d = 1.50 \text{ in}, \ d = 1.75 \text{ in}, \ d = 2 \text{ in}, \\ 1/3 \text{ in} & \text{for } d = 2.5 \text{ in}, \\ 1/2 \text{ in} & \text{for } d = 3 \text{ in}. \end{cases} \tag{3.18}$$

3.3. Threaded fasteners

An external force F_e tends to pull apart the two parts connected with a bolt as shown in Fig. 3.6. An initial preload F_i acts on the bolt and plates

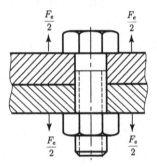

Figure 3.6 Parts connected with a bolt.

$$F_i = F_{b0} = F_{c0},$$

where F_{b0} is the initial bolt axial force and F_{c0} is initial clamping force between the parts. The external separating force F_e is

$$F_e = \Delta F_b + \Delta F_c, \tag{3.19}$$

where ΔF_b is the increase in bolt force and ΔF_c is the decrease in clamping force. The elastic displacements of the bolt and clamped parts are equal

$$\delta = \frac{\Delta F_b}{k_b} = \frac{\Delta F_c}{k_c}, \tag{3.20}$$

where k_b is the spring constant for the bolt and k_c is the spring constant for the clamped parts. The elongation is calculated from Eqs. (3.19) and (3.20) as

$$\delta = \frac{F_e}{k_b + k_c}. \tag{3.21}$$

The bolt axial force F_b is

$$F_b = F_i + \Delta F_b = F_i + \frac{k_b}{k_b + k_c} F_e. \tag{3.22}$$

The clamping force F_c is

$$F_c = F_i - \Delta F_c = F_i - \frac{k_c}{k_b + k_c} F_e. \tag{3.23}$$

The *joint constant* is defined as

$$C = \frac{k_b}{k_b + k_c}. \tag{3.24}$$

The bolt axial and clamping forces, calculated in terms of the joint constant, are

$$F_b = F_i + C F_e, \tag{3.25}$$
$$F_c = F_i - (1 - C) F_e. \tag{3.26}$$

The general expression for the spring constant (stiffness) is

$$k = \frac{F}{\delta} = F \frac{A E}{F l} = \frac{A E}{l},$$

where A is the cross-sectional area, E is the modulus of elasticity, and l is the length.

Bolt stiffness

The spring constant (stiffness) of a bolt is calculated with

$$\frac{1}{k_b} = \frac{1}{k_s} + \frac{1}{k_t},$$

where k_s is the stiffness of the unthreaded section of the bolt (shank) and k_t is the stiffness of the threaded section of the bolt. The stiffness of the unthreaded section of the bolt is [29]

$$k_s = \frac{A_s E}{l_s} = \frac{\pi d^2 E}{4 l_s}, \tag{3.27}$$

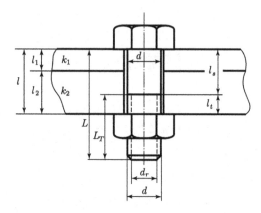

Figure 3.7 Threaded fastener.

where d is the major diameter and l_s is the length of the unthreaded section in grip as shown in Fig. 3.7. The stiffness of the threaded section of the bolt is [29]

$$k_t = \frac{A_t E}{l_t},$$

(3.28)

where l_t is the length of the threaded section in grip as shown in Fig. 3.7 and A_t is the *tensile stress area*. The stiffness of the bolt is

$$\frac{1}{k_b} = \frac{1}{k_t} + \frac{1}{k_s} = \frac{l_t}{A_t E} + \frac{l_s}{A_s E},$$

(3.29)

or

$$k_b = \frac{A_s A_t E}{A_s l_t + A_t l_s}.$$

(3.30)

The tensile stress area, A_t, is defined as [9]:

$$A_t = 0.7854 \left(d - 0.9743/n\right)^2 \text{ in}^2,$$

(3.31)

for UN thread profiles, where d is in inches and n is the number of threads per inch, and

$$A_t = 0.7854(d - 0.9382p)^2 \text{ mm}^2,$$

(3.32)

for M thread profiles with the major diameter d and the pitch p in millimeters.

For UN screws the relation between the major diameter, d, and the number of threads per inch, n, is [9]:

- for coarse threads (UNC):

$$
d = \begin{cases}
0.125 \text{ in} & \text{the number of threads per inch is } n = 40, \\
0.25 \text{ in} & \text{the number of threads per inch is } n = 24, \\
0.5 \text{ in} & \text{the number of threads per inch is } n = 13, \\
0.75 \text{ in} & \text{the number of threads per inch is } n = 10, \\
1 \text{ in} & \text{the number of threads per inch is } n = 8, \\
1.125 \text{ in} & \text{the number of threads per inch is } n = 7, \\
1.25 \text{ in} & \text{the number of threads per inch is } n = 7, \\
1.5 \text{ in} & \text{the number of threads per inch is } n = 6, \\
1.75 \text{ in} & \text{the number of threads per inch is } n = 5, \\
2 \text{ in} & \text{the number of threads per inch is } n = 4.5,
\end{cases} \tag{3.33}
$$

- for fine threads (UNF):

$$
d = \begin{cases}
0.125 \text{ in} & \text{the number of threads per inch is } n = 44, \\
0.25 \text{ in} & \text{the number of threads per inch is } n = 28, \\
0.5 \text{ in} & \text{the number of threads per inch is } n = 20, \\
0.75 \text{ in} & \text{the number of threads per inch is } n = 16, \\
1 \text{ in} & \text{the number of threads per inch is } n = 12, \\
1.125 \text{ in} & \text{the number of threads per inch is } n = 12, \\
1.25 \text{ in} & \text{the number of threads per inch is } n = 12, \\
1.5 \text{ in} & \text{the number of threads per inch is } n = 12.
\end{cases} \tag{3.34}
$$

The tensile stress can be calculated using the major diameter, d, and the pitch, p, and the relation between the major diameter and pitch is given by [9]:

- for coarse threads (MC):

$$
d = \begin{cases}
1 \text{ mm} & \text{the pitch is } p = 0.25 \text{ mm,} \\
2 \text{ mm} & \text{the pitch is } p = 0.4 \text{ mm,} \\
2.5 \text{ mm} & \text{the pitch is } p = 0.45 \text{ mm,} \\
3 \text{ mm} & \text{the pitch is } p = 0.5 \text{ mm,} \\
4 \text{ mm} & \text{the pitch is } p = 0.7 \text{ mm,} \\
5 \text{ mm} & \text{the pitch is } p = 0.8 \text{ mm,} \\
6 \text{ mm} & \text{the pitch is } p = 1 \text{ mm,} \\
8 \text{ mm} & \text{the pitch is } p = 1.25 \text{ mm,} \\
10 \text{ mm} & \text{the pitch is } p = 1.5 \text{ mm,} \\
12 \text{ mm} & \text{the pitch is } p = 1.75 \text{ mm,} \\
16 \text{ mm} & \text{the pitch is } p = 2 \text{ mm,} \\
20 \text{ mm} & \text{the pitch is } p = 2.5 \text{ mm,} \\
24 \text{ mm} & \text{the pitch is } p = 3 \text{ mm,}
\end{cases} \tag{3.35}
$$

- for fine threads (MF):

$$
d = \begin{cases}
2 \text{ mm} & \text{the pitch is } p = 0.25 \text{ mm,} \\
2.5 \text{ mm} & \text{the pitch is } p = 0.35 \text{ mm,} \\
3 \text{ mm} & \text{the pitch is } p = 0.35 \text{ mm,} \\
4 \text{ mm} & \text{the pitch is } p = 0.5 \text{ mm,} \\
5 \text{ mm} & \text{the pitch is } p = 0.5 \text{ mm,} \\
6 \text{ mm} & \text{the pitch is } p = 0.75 \text{ mm,} \\
8 \text{ mm} & \text{the pitch is } p = 1 \text{ mm,} \\
10 \text{ mm} & \text{the pitch is } p = 1.25 \text{ mm,} \\
12 \text{ mm} & \text{the pitch is } p = 1.25 \text{ mm,} \\
16 \text{ mm} & \text{the pitch is } p = 1.5 \text{ mm,} \\
20 \text{ mm} & \text{the pitch is } p = 1.5 \text{ mm,} \\
24 \text{ mm} & \text{the pitch is } p = 2 \text{ mm.}
\end{cases} \tag{3.36}
$$

The total threaded length L_T, as shown in Fig. 3.7, is calculated with [4]

$$
L_T \text{ (in)} = \begin{cases}
2d + 1/4 \text{ in} & \text{for } L \le 6 \text{ in,} \\
2d + 1/2 \text{ in} & \text{for } L > 6 \text{ in,}
\end{cases} \tag{3.37}
$$

and

$$L_T \text{ (mm)} = \begin{cases} 2d + 6 \text{ mm} & \text{for } L \leq 125 \text{ mm and } d \leq 48 \text{ mm,} \\ 2d + 12 \text{ mm} & \text{for } 125 < L \leq 2000 \text{ mm,} \\ 2d + 25 \text{ mm} & \text{for } L > 2000 \text{ mm,} \end{cases} \tag{3.38}$$

where L is the total length of the bolt.

Stiffness of the clamped parts

Estimating the stiffness of the clamped parts is a difficult task. The total clamped part is a combination of elements made of the same or different materials.

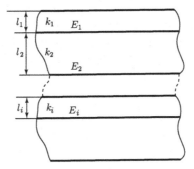

Figure 3.8 Clamped parts.

The parts may be considered as "springs" in series, as shown in Fig. 3.8, and the stiffness of the clamped part is

$$\frac{1}{k_c} = \frac{1}{k_1} + \frac{1}{k_2} + \cdots + \frac{1}{k_i} + \cdots .$$

For parts with the same modulus of elasticity, E, the stiffness of the clamped parts is [29]

$$k_m = \frac{0.577\pi Ed}{2\ln\left(5\dfrac{0.577l + 0.5d}{0.577l + 2.5d}\right)}, \tag{3.39}$$

where l is the length of the grip. Wileman et al. [33] obtained an exponential expression using finite element analysis to compute the stiffness of the clamped part with the same modulus of elasticity

$$k_m = EdA e^{Bd/l}, \tag{3.40}$$

where the numerical constants are:

$A = 0.78715,$ $\quad B = 0.62873$ for steel ($E = 206.8$ GPa),

$A = 0.79670,$ $\quad B = 0.63816$ for aluminum ($E = 71$ GPa),

$A = 0.79568,$ $\quad B = 0.63553$ for copper ($E = 118.6$ GPa),

$A = 0.77871,$ $\quad B = 0.61616$ for gray cast iron ($E = 100$ GPa), and

$A = 0.78952,$ $\quad B = 0.62914$ for a general metal.

The clamped parts may have distinct material properties with the Young modulus E_i and thicknesses l_i, as shown in Fig. 3.8. The values of A and B are taken from the general case: $A = 0.78952$ and $B = 0.62914$. The equivalent member modulus, E, is calculated with [5]

$$\frac{1}{E} = \frac{1}{l} \sum_{i=1}^{n} \frac{l_i}{E_i} \quad \text{where} \quad l = \sum_{i=1}^{n} l_i. \tag{3.41}$$

Bolt preload

The initial tensile force F_i is calculated as [11]:

$$F_i = KA_t S_p, \tag{3.42}$$

where K is a preload factor, A_t is the tensile stress area of the thread and S_p is the *proof strength* of the material [11,29].

The International Organization for Standardization (ISO) describes a metric grade number is a range from 4.6 to 12.9, and the proof strength can be determined as [9]

$$S_p = \begin{cases} 225 \text{ MPa} & \text{for } \mathbf{M} \text{ grade } \mathbf{4.6} & \text{and M 5–M 36,} \\ 310 \text{ MPa} & \text{for } \mathbf{M} \text{ grade } \mathbf{4.8} & \text{and M 1.6–M 16,} \\ 380 \text{ MPa} & \text{for } \mathbf{M} \text{ grade } \mathbf{5.8} & \text{and M 5–M 24,} \\ 600 \text{ MPa} & \text{for } \mathbf{M} \text{ grade } \mathbf{8.8} & \text{and M 17–M 36,} \\ 650 \text{ MPa} & \text{for } \mathbf{M} \text{ grade } \mathbf{9.8} & \text{and M 1.6–M 16,} \\ 830 \text{ MPa} & \text{for } \mathbf{M} \text{ grade } \mathbf{10.9} & \text{and M 16–M 36,} \\ 970 \text{ MPa} & \text{for } \mathbf{M} \text{ grade } \mathbf{12.9} & \text{and M 1.6–M 36.} \end{cases} \tag{3.43}$$

The Society of Automotive Engineers (SAE) specifies the grade number from 1 to 8, and the proof strength is calculated with [9]

$$S_p = \begin{cases} 33 \text{ kpsi} & \text{for } \textbf{SAE} \text{ grade } \textbf{1} & \text{and } d = 0.25\text{--}1.5 \text{ in,} \\ 55 \text{ kpsi} & \text{for } \textbf{SAE} \text{ grade } \textbf{2} & \text{and } d = 0.25\text{--}0.75 \text{ in,} \\ 33 \text{ kpsi} & \text{for } \textbf{SAE} \text{ grade } \textbf{2} & \text{and } d = 0.75\text{--}1.5 \text{ in,} \\ 65 \text{ kpsi} & \text{for } \textbf{SAE} \text{ grade } \textbf{4} & \text{and } d = 0.25\text{--}1.5 \text{ in,} \\ 85 \text{ kpsi} & \text{for } \textbf{SAE} \text{ grade } \textbf{5} & \text{and } d = 0.25\text{--}1 \text{ in,} \\ 74 \text{ kpsi} & \text{for } \textbf{SAE} \text{ grade } \textbf{5} & \text{and } d = 1\text{--}1.5 \text{ in,} \\ 105 \text{ kpsi} & \text{for } \textbf{SAE} \text{ grade } \textbf{7} & \text{and } d = 0.25\text{--}1.5 \text{ in,} \\ 120 \text{ kpsi} & \text{for } \textbf{SAE} \text{ grade } \textbf{8} & \text{and } d = 0.25 - 1.5 \text{ in.} \end{cases} \tag{3.44}$$

The preload factor K is 0.75 for reused connections and 0.90 for permanent connections. The *proof load* is defined as

$$F_p = A_t S_p, \tag{3.45}$$

and the torque necessary to achieve the preload is

$$T = K_T F_i d, \tag{3.46}$$

where K_T is a torque factor given by

$$K_T = \begin{cases} 0.30 & \text{for non-plated, black finish,} \\ 0.20 & \text{for zinc-plated,} \\ 0.18 & \text{for lubricated,} \\ 0.16 & \text{for cadmium-plated.} \end{cases} \tag{3.47}$$

Static loading of the joint
The bolt stress can be calculated from Eq. (3.25) as

$$\sigma_b = \frac{F_b}{A_t} = \frac{F_i}{A_t} + C\frac{F_e}{A_t}, \tag{3.48}$$

where A_t is the tensile stress area. The limiting value for the bolt stress, σ_b, is the proof strength, S_p. A safety factor n_y is introduced for yielding, and then Eq. (3.48) becomes

$$S_p = \frac{n_y F_b}{A_t} \quad \text{and} \quad n_y = \frac{S_p A_t}{C F_e + F_i}. \tag{3.49}$$

A load factor, n_b, for the bolt against the external load is introduced [4] as

$$S_p = \frac{F_i}{A_t} + C\frac{n_b F_e}{A_t} \quad \text{and} \quad n_b = \frac{S_p A_t - F_i}{C F_e}. \tag{3.50}$$

The load factor is not applied to the preload stress F_i/A_t. If $n_b > 1$, the bolt stress is less than the proof strength. When the clamping force is zero, $F_c = 0$, there will be a detachment of the parts. The safety factor against detachment of the parts of the joint is obtained from Eq. (3.26) with $F_c = 0$ and has the expression

$$n_s = \frac{F_i}{F_e(1 - C)}. \tag{3.51}$$

3.4. Examples

Example 3.1. The scissor jack shown in Fig. 3.9 is subjected to a vertical force $P = 10$ kN. The square-threaded screw 1 has a major diameter $d = 38$ mm and pitch $p = 6$ mm. The thread coefficient of friction at E is $\mu_s = 0.2$, and the angle of the jack is $\theta = 30°$. Determine (a) the moment to raise and lower the load; (b) if the screw is self-locking; (c) the starting moment required to raise the load.

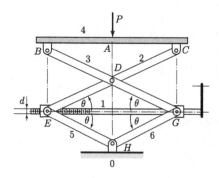

Figure 3.9 Example 3.1.

Solution

(a) The axial force F that acts on screw 1 has to be calculated. The force diagram for link 4 is shown in Fig. 3.10A. The force of link 3 on link 4 is F_{34} and is determined from the sum of the moments on link 4 with respect to joint C:

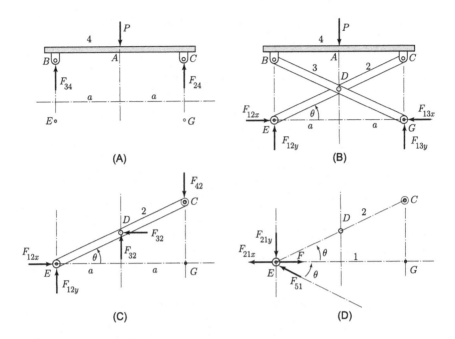

Figure 3.10 Free body diagrams of Example 3.1.

$$-F_{34}\,a + P\,a = 0 \quad \text{or} \quad F_{34} = P/2.$$

The force of link 2 on link 4 is F_{24} and is determined from the sum of the forces on link 4:

$$F_{34} + F_{24} + P = 0 \quad \text{or} \quad F_{24} = P/2.$$

The free body diagram of links 2, 3, and 4 is shown in Fig. 3.10B. The sum of the moments with respect to joint G gives

$$-F_{12y}\,(2\,a) + P\,a = 0,$$

and the reaction force of screw 1 on link 2 along the y-axis at E is

$$F_{12y} = P/2.$$

The free body diagram for link 2 is shown in Fig. 3.10C. The sum of the moments on link 2 with respect to joint D gives

$$F_{12x}\,a\tan\theta - \left(F_{42} + F_{12y}\right)a = 0,$$

where the reaction force of link 4 on link 2 is $F_{42} = F_{24}$ and the reaction force of link 1 on link 2 along the x-axis is

$$F_{12x} = (F_{42} + F_{12y}) / \tan \theta.$$

The free body diagram of node E is shown in Fig. 3.10D. The components of the force of link 2 on screw 1 are $F_{21x} = F_{12x}$ and $F_{21y} = F_{12y}$. Link 5 is a two-force member, and the force of link 5 on screw 1, F_{51}, is along EH. The sum of the forces on the y-axis at node E is

$$-F_{21y} + F_{51} \sin \theta = 0 \quad \text{or} \quad F_{51} = F_{21y} / \sin \theta.$$

The axial force F that acts on screw 1 is calculated from the sum of the forces on the x-axis

$$F = F_{21x} + F_{51} \cos \theta.$$

The MATLAB commands and numerical results are:

```
P =  10;        % kN external load
theta = pi/6;   % rad
F34 = P/2;
F24 = P/2;
F42 = F24;
F12y = P/2;
F12x = (F42+F12y)/tan(theta);
F21x = F12x;
F21y = F12y;
F51 = F21y/sin(theta);
F = F21x + F51*cos(theta);
% F34 =   5.000 (kN)
% F24 =   5.000 (kN)
% F12y =  5.000 (kN)
% F12x = 17.321 (kN)
% F51 = 10.000 (kN)
% F = 25.981 (kN)
```

From Fig. 3.3 the minor diameter is $d_r = d - p$, while the pitch (mean) diameter is $d_m = d - p/2$. The lead of the screw is $l = p$, the lead angle is $\lambda = \arctan [l/(\pi d_m)]$ and the friction angle is $\theta_s = \arctan \mu$. The MATLAB commands and numerical results are:

```
d = 38*10^(-3);    % (m) major diameter
p = 6*10^(-3);     % (m) pitch
mu = 0.2;          % coefficient of friction
dr = d - p;
dm = d - p/2;
l = p;
lambda = atan(l/(pi*dm));
thetas = atan(mu);
% dr =  0.032 (m)
% dm =  0.035 (m)
% l =  0.006 (m)
% lambda =  3.123 (deg)
% thetas = 11.310 (deg)
```

The moment required to overcome the thread friction and to raise the load is

$$M_r = \frac{F d_m}{2}\left(\frac{l + \pi \mu d_m}{\pi d_m - \mu l}\right),$$

the moment required to lower the load is

$$M_l = \frac{F d_m}{2}\left(\frac{\pi \mu d_m - l}{\pi d_m + \mu l}\right),$$

or with MATLAB:

```
Mr = 0.5*F*dm*(pi*mu*dm+l)/(pi*dm-mu*l);
Ml = 0.5*F*dm*(pi*mu*dm-l)/(pi*dm+mu*l);
% Mr = 117.020 (N m)
% Ml = 65.409 (N m)
```

(b) The condition for self-locking is

$$\pi \mu d_m - l > 0,$$

and with MATLAB:

```
sf = (pi*mu*dm - l);
if sf > 0
fprintf('sf>0 => screw is self-locking\n\n')
else
fprintf('sf<0 => screw is not self-locking\n\n')
```

```
end
% sf =   0.016
% sf>0 => screw is self-locking
```

The screw is self-locking and it will not unscrew even if the force is re-moved.

(c) The starting friction is about one-third higher than running fric-tion [11], and the coefficient of starting friction is $\mu_s = \dfrac{4}{3}\mu$. The starting moment for lifting the load is:

```
mus = 4*mu/3;
Mrs = 0.5*F*dm*(pi*mus*dm+l)/(pi*dm-mus*l);
% mus =   0.267
% Mrs = 148.210 (N m)
```

Example 3.2. The car jack is made from a threaded shaft 1 and four pinned members 2, 3, 4, and 5 as shown in Fig. 3.11A. The links of the jack are pinned together at points A, B, C, and D. The length of the links is $AB = BC = CD = DA = 8$ in. The threaded shaft 1 is pinned to the jack at points A and C. Each single square-threaded screw has a major diameter of $d = 1.125$ in and 7 square threads per in. The thread coefficient of friction at A and C is $\mu = 0.3$ and the angle of the jack is $\theta = 45°$. The external weight exerts a force of $P = 400$ lb on the jack. Determine the horizontal

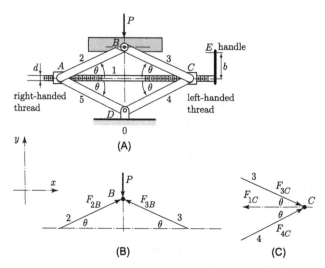

Figure 3.11 Example 3.2.

forces that must be applied perpendicular to the handle at E to raise and lower the load if $b = 4$ in.

Solution

The force in the threaded shaft 1 can be obtained by analyzing the free body diagrams of joints B and C. The free body diagram of joint B is shown in Fig. 3.11B. The equilibrium of joint B gives

$$\sum F_x = F_{2B} \cos \theta - F_{3B} \cos \theta = 0,$$

and the force of link 2 on joint B is equal to the force of link 3 on joint B, that is,

$$F_{2B} = F_{3B}.$$

The sum of the forces on the y-axis at B is

$$\sum F_y = F_{2B} \sin \theta + F_{3B} \sin \theta - P = 0,$$

resulting in

$$F_{2B} = F_{3B} = 0.5P / \sin \theta.$$

Link 3 is a two-force member and

$$F_{3C} = F_{3B}.$$

The free body diagram of joint C is shown in Fig. 3.11C, and the sum of the forces on the y-axis is

$$\sum F_y = -F_{3C} \sin \theta + F_{4C} \sin \theta = 0,$$

or the force of link 3 on joint C is equal to the force of link 4 on joint C, i.e.,

$$F_{4C} = F_{3C}.$$

The sum of the forces on the x-axis is

$$\sum F_x = F_{3C} \cos \theta + F_{4C} \cos \theta = F_{1C}.$$

The force on the threaded shaft 1 is

$$F = F_{1C} = 2F_{3C} \cos \theta = 2(0.5P / \sin \theta) \cos \theta = P / \tan \theta.$$

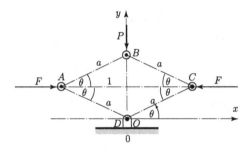

Figure 3.12 Virtual work of Example 3.2.

Another approach to derive an equation for the tension F is to determine the virtual work performed by the weight $\mathbf{P} = -P\mathbf{J}$ and the force on the threaded shaft 1. Knowing that the total virtual work is equal to zero, tension F is derived as a function of angle θ. The free-body diagram of the jack is shown in Fig. 3.12. Point D is fixed as the origin of the system, and the position vector of point B is

$$\mathbf{r}_B = x_B \mathbf{1} + y_B \mathbf{J} = 2a\sin\theta\,\mathbf{J}.$$

The virtual displacement of point B is

$$\delta\mathbf{r}_B = 2a\cos\theta\,\delta\theta\,\mathbf{J}.$$

The virtual work performed by weight P is

$$\delta U_P = \mathbf{P}\cdot\delta\mathbf{r}_B = -2Pa\cos\theta\,\delta\theta.$$

The tension in the threaded shaft can be divided into two equal and opposite horizontal components, $\mathbf{F}_A = F\mathbf{1}$ at A and $\mathbf{F}_C = -F\mathbf{1}$ at C. The position vector of point A is

$$\mathbf{r}_A = -a\cos\theta\,\mathbf{1} + a\sin\theta\,\mathbf{J},$$

and the virtual displacement of point A is

$$\delta\mathbf{r}_A = a\sin\theta\,\delta\theta\,\mathbf{1} + a\cos\theta\,\delta\theta\,\mathbf{J}.$$

The position vector of point C is

$$\mathbf{r}_C = a\cos\theta\,\mathbf{1} + a\sin\theta\,\mathbf{J},$$

and the virtual displacement of point C is

$$\delta \mathbf{r}_C = -a \sin \theta \, \delta \theta \, \mathbf{1} + a \cos \theta \, \delta \theta \, \mathbf{J}.$$

The virtual works performed by the two components of tension, $\mathbf{F}_A$ and $\mathbf{F}_C$, are

$$\delta U_{F_A} = \mathbf{F}_A \cdot \delta \mathbf{r}_A = Fa \sin \theta \, \delta \theta \quad \text{and} \quad \delta U_{F_C} = \mathbf{F}_C \cdot \delta \mathbf{r}_C = Fa \sin \theta \, \delta \theta.$$

The total virtual work is the sum of the virtual work performed by the two components of the tension, $\mathbf{F}_A$ and $\mathbf{F}_C$, and the virtual work performed by the weight $\mathbf{P}$, namely

$$\delta U = \delta U_{F_A} + \delta U_{F_C} + \delta U_P = Fa \sin \theta \, \delta \theta + Fa \sin \theta \, \delta \theta - 2Pa \cos \theta \, \delta \theta.$$

The principle of virtual work states that a system of rigid bodies is in equilibrium if the total virtual work done by all external forces and couples acting on the system is zero for each independent virtual displacement of the system

$$\delta U = 0,$$

or

$$F = P / \tan \theta.$$

The MATLAB program for the principle of virtual work is:

```
% virtual work method
syms a theta P F
rB_ = [0, 2*a*sin(theta), 0];
% virtual displacements drB_
drB_ = diff(rB_,theta);
P_ = [0,-P,0];
dUP = P_*drB_.'; % virtual work
rA_ = [-a*cos(theta), a*sin(theta), 0];
% virtual displacements drA_
drA_ = diff(rA_,theta);
FA_ = [F, 0, 0];
dUFA = FA_*drA_.'; % virtual work
rC_ = [a*cos(theta), a*sin(theta), 0];
% virtual displacements drC_
drC_ = diff(rC_,theta);
```

```
FC_ = [-F, 0, 0];
dUFC = FC_*drC_.'; % virtual work
% total virtual work
dU = dUP+dUFA+dUFC;
Fls = solve(dU,F);
% F = (P*cos(theta))/sin(theta)
figure(1)
thet = 0:1:90;
plot(thet, 400./tand(thet), 'b-','LineWidth',2)
xlabel('\theta (degree)','FontSize',14)
ylabel('F (lb)','FontSize',14)
grid
```

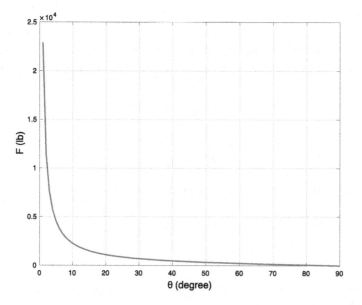

Figure 3.13 Force F versus angle θ of Example 3.2.

The graph of force F versus angle θ is shown in Fig. 3.13. Tension F increases as angle θ approaches 0. The normal operating conditions are $30° \leq \theta \leq 60°$. The pitch p, mean diameter d_m, lead l, lead angle λ, and friction angle θ_s are calculated with:

```
d = 1.125; % (in) major diameter
mu = 0.3;  % coefficient of friction
% 7 square threads per in
```

```
p = 1/7;        % (in) pitch
dm = d-p/2;     % mean diameter
l = p; % lead single square-threaded
lambda = atan(l/(pi*dm)); % lead angle
thetas = atan(mu); % friction angle
```

and the numerical results are:

```
% p =   0.143 (in)
% dm =   1.054 (in)
% l =   0.143 (in)
% lambda =   2.471 (deg)
% thetas = 16.699 (deg)
```

The moments required to raise and lower the load for the two screws are:

```
Mr = F*dm*(pi*mu*dm+l)/(pi*dm-mu*l);
% Mr = 146.515 (lb in)
Ml = F*dm*(pi*mu*dm-l)/(pi*dm+mu*l);
% Ml = 106.856 (lb in)
```

The same moments can be calculated with the expressions:

```
%  moment required to raise the load
Mr = 2*F*(dm/2)*tan(thetas+lambda);
%  moment required to lower the load
Ml = 2*F*(dm/2)*tan(thetas-lambda);
```

The horizontal forces that must be applied perpendicular to the handle at E to raise and to lower the load are:

```
Fr = Mr/b;
Fl = Ml/b;
% Fr = 36.629 (lb)
% Fl = 26.714 (lb)
```

Example 3.3. The mechanism shown in Fig. 3.14 consists of link 2 that has a double Acme-threaded screw with a major diameter $d = 1.5$ in, pitch $p = 1/4$ in and coefficient of friction for the thread $\mu = 0.2$. Load $P = 2000$ lb is acting at the end of link 1 at B. The length of link 2 is $BD = L = 45$ in and the angle of link 2 with the horizontal is $\theta = \pi/6$ rad. The following dimensions are given: $OD = L$ and $BC = CD = L/2$. Determine (a) the starting moment to raise and lower the load using the

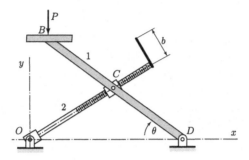

Figure 3.14 Example 3.3.

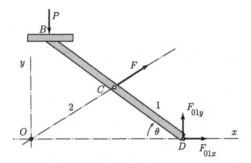

Figure 3.15 Free body diagram of link 1 in Example 3.1.

coefficient of starting friction; (b) the moment to raise and lower the load; (c) if the screw is self-locking; (d) the efficiency of the screw; (e) plot the efficiency of the screw as a function of angle λ for different coefficients of friction.

Solution

(a) The force on the threaded shaft 2 is obtained using the free body diagrams of link 1 as shown in Fig. 3.15. The coordinates of points D, B, and C are determined with:

```
L = 45;         % in
theta = pi/6; % rad
x0 = 0; y0 = 0;
xD = L; yD = 0;
xB = xD - L*cos(theta);
yB = L*sin(theta);
xC = xD - 0.5*L*cos(theta);
yC = 0.5*L*sin(theta);
```

```
r0_ = [x0 y0 0];
rD_ = [xD yD 0]; % position vector of D
rB_ = [xB yB 0]; % position vector of B
rC_ = [xC yC 0]; % position vector of C
% rD_ = [45.000  0.000  0.000] (in)
% rB_ = [ 6.029 22.500  0.000] (in)
% rC_ = [25.514 11.250  0.000] (in)
```

The unit vector for the direction of the force, *F*, on the threaded shaft 2 is:

```
uF_ = rC_/norm(rC_);
% uF_ = [ 0.915  0.403  0.000]
```

and the symbolic force on the threaded shaft is:

```
syms F
F_ = F*uF_;
```

The force *F* is calculated using the sum of the moments on link 1 about the point *D*:

```
P = -2000;    % lb
P_ = [0 P 0]; % force at B
% sum of moments on link 1 about D
% M1D_ = cross(rC_-rD_, F_) + cross(rB_-rD_, P_);
M1D_ = cross(rC_-rD_, F_) + cross(rB_-rD_, P_);
F = eval(solve(M1D_(3),F));
% F = 4293.109 (lb)
```

The graph of the jack (see Fig. 3.16) and the vector forces are obtained in MATLAB with:

```
a = L;
axis([-a a -a a])
grid on, hold on
xlabel('x'), ylabel('y')

line([x0 xC],[y0 yC],'LineStyle','--',...
    'Color','r','LineWidth',2)
line([xD xB],[y0 yB],'LineStyle','-',...
    'Color','k','LineWidth',3)
```

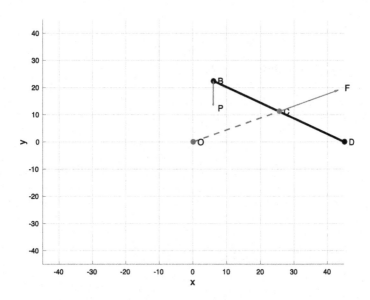

Figure 3.16 Graph of the jack in Example 3.3.

```
text(x0,y0,0,'  0','fontsize',12);
text(xB,yB,0,'  B','fontsize',12);
text(xC,yC,0,'  C','fontsize',12);
text(xD,yD,0,'  D','fontsize',12);

plot(x0,y0,'ro',...
    'MarkerSize',6,'LineWidth',2,'MarkerFaceColor','r')
plot(xB,yB,'ko',...
    'MarkerSize',6,'LineWidth',2,'MarkerFaceColor','k')
plot(xC,yC,'bo',...
    'MarkerSize',6,'LineWidth',2,'MarkerFaceColor','b')
plot(xD,yD,'ko',...
    'MarkerSize',6,'LineWidth',2,'MarkerFaceColor','k')

af = 200; % force factor for graph
quiver(xB,yB,0,P/af,...
'Color','b','LineWidth',1.5);
quiver(xC,yC,F*uF_(1)/af,F*uF_(2)/af,...
'Color','b','LineWidth',1.5);
text(xB,yB+P/af,0,'  P','fontsize',12);
text(xC+F*uF_(1)/af,yC+F*uF_(2)/af,0,'F','fontsize',12);
```

For the major diameter $d = 1.5$ in the preferred screw, the pitch is $p = 0.25$ in. Because of the double thread, the lead is $l = 2p$, the pitch (mean) diameter is $d_m = d - p/2$, and the helix angle is $\lambda = \arctan \dfrac{l}{\pi\, d_m}$. The MATLAB results are:

```
d = 1.5;  % (in)
p = 1/4; % (in)
l = 2*p;
dm = d - p/2;
lambda = atan(l/(pi*dm));
% p =   0.250 (in)
% l =   0.500 (in)
% dm =  1.375 (in)
% lambda =  6.603 (deg)
```

For the Acme threads, the thread angle $\alpha = 14.5°$ (see Fig. 3.3), and the thread angle measured in normal plane is $\tan\alpha_n = \tan\alpha\cos\lambda$, or with MATLAB:

```
alpha = (29/2)*(pi/180);
alphan = atan(tan(alpha)*cos(lambda));
% alphan = 14.408 (deg)
```

The starting friction is about one-third higher than running friction, and the coefficient of starting friction is $\mu_s = \dfrac{4}{3}\mu$. The starting moments for lifting and for lowering the load are:

```
mus = (4/3)*mu;
Mrs=0.5*F*dm*(pi*mus*dm+l*cos(alphan))/(pi*dm*cos(alphan)-mus*l);
Mls=0.5*F*dm*(pi*mus*dm-l*cos(alphan))/(pi*dm*cos(alphan)+mus*l);
% Mrs = 1192.258 (lb in)
% Mls = 456.447 (lb in)
```

(b) Changing the coefficient of friction to the running value of μ, the moments for lifting and for lowering the load with running friction are:

```
Mr=0.5*F*dm*(pi*mu*dm+l*cos(alphan))/(pi*dm*cos(alphan)-mu*l);
Ml=0.5*F*dm*(pi*mu*dm-l*cos(alphan))/(pi*dm*cos(alphan)+mu*l);
% Mr = 974.395 (lb in)
% Ml = 261.584 (lb in)
```

(c) The self-locking condition is

```
sf = (pi*mu*dm-l*cos(alphan));
if sf > 0
fprintf('sf>0 => screw is self-locking\n\n')
else
fprintf('sf<0 => screw is not self-locking\n\n')
end
% sf =   0.380
```

and the screw is not self-locking.

(d) With friction coefficient being zero, the moment to raise the load is

$$M_0 = \frac{F d_m}{2} \left(\frac{l \cos \alpha_n}{\pi d_m \cos \alpha_n} \right) = \frac{Fl}{2\pi}.$$

The efficiency is the ratio of friction-free moment to actual moment, or

$$e = \frac{Fl}{2\pi M_r} = \frac{M_0}{M_r}.$$

The MATLAB calculations for the efficiency are:

```
% moment for raising the load with no friction
Mr0 = 0.5*F*dm*(l*cos(alphan))/(pi*dm*cos(alphan)); % or
Mr0 = F*l/(2*pi);
% efficiency: ratio of friction-free moment to actual moment
e = Mr0/Mr;
% Mr0 = 341.635 (lb in)
% e =   0.351
```

(e) The efficiency of the screw can be calculated with

$$e = \frac{\cos \alpha_n - \mu \tan \lambda}{\cos \alpha_n + \mu \cot \lambda}.$$

The efficiency is a function of λ and is plotted for different μ. The results are shown in Fig. 3.17. The MATLAB program for the efficiency function is:

```
lambdae = 0:pi/180:90*pi/180;
alphane = atan(tan(alpha)*cos(lambdae));
```

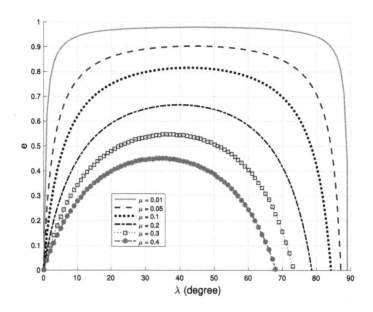

Figure 3.17 Efficiency *e* as a function of λ in Example 3.3.

```
mu01 = 0.01;
e101 = (cos(alphane)-mu01*tan(lambdae));
e201 = (cos(alphane)+mu01*cot(lambdae));
ee01 = e101./e201;

mu05 = 0.05;
e105 = (cos(alphane)-mu05*tan(lambdae));
e205 = (cos(alphane)+mu05*cot(lambdae));
ee05 = e105./e205;

mu1 = 0.1;
e11 = (cos(alphane)-mu1*tan(lambdae));
e21 = (cos(alphane)+mu1*cot(lambdae));
ee1 = e11./e21;

mu2 = 0.2;
e12 = (cos(alphane)-mu2*tan(lambdae));
e22 = (cos(alphane)+mu2*cot(lambdae));
ee2 = e12./e22;
```

```
mu3 = 0.3;
e13 = (cos(alphane)-mu3*tan(lambdae));
e23 = (cos(alphane)+mu3*cot(lambdae));
ee3 = e13./e23;

mu4 = 0.4;
e14 = (cos(alphane)-mu4*tan(lambdae));
e24 = (cos(alphane)+mu4*cot(lambdae));
ee4 = e14./e24;

hold on
plot(lambdae*180/pi, ee01,'-','LineWidth',2)
plot(lambdae*180/pi, ee05,'--b','LineWidth',2)
plot(lambdae*180/pi, ee1,':k','LineWidth',3)
plot(lambdae*180/pi, ee2,'-.k','LineWidth',2)
plot(lambdae*180/pi, ee3,':bs','LineWidth',1)
plot(lambdae*180/pi, ee4,'-.r*','LineWidth',2)

xlabel('\lambda (degree)','FontSize',14)
ylabel('e','FontSize',14)
axis([0 90 0 1 ])
grid on
legend(....
'\mu = 0.01','\mu = 0.05',...
'\mu = 0.1','\mu = 0.2',...
'\mu = 0.3','\mu = 0.4')
```

Example 3.4. For a joint, a steel bolt 0.5–13 UNC of SAE class 4 is selected. The clamp length is $l_m = 3$ in, and the total bolt length $L = 3.5$ in as shown in Fig. 3.18. The Young's modulus of all the elements is $E = 30\,(10^6)$ psi. An external force of $F_e = 2500$ lb acts on the joint and tends to separate the two parts. Find the bolt load factor and the safety factor against joint separation for reused and permanent connections. Plot the bolt load and safety factors against joint separation as a function of preload factor K defined by Eq. (3.42) [9].

Solution

A bolt 0.5–13 UNC has the major diameter $d = 0.5$ in and has $n = 13$ threads/inch. The shank area is calculated with:

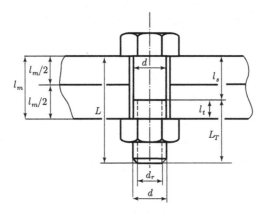

Figure 3.18 Example 3.4.

```
As = (pi * d^2) / 4;
% As = 0.196 (in^2)
```

The proof strength of the steel bolt SAE 4 with $d = 0.5$ in is given by Eq. (3.44), and it is $S_p = 65$ kpsi. The tensile stress area of the thread is calculated with Eq. (3.31):

```
At = (pi/4)*(d-0.9743/n)^2;
% At =  0.142 (in^2)
```

The length of the thread of the bolt, L_T, is given by Eq. (3.37):

```
LT = 2*d + 0.25;
% LT =  1.250 (in)
```

The length of the shank of the bolt, l_s, is:

```
ls = L - LT;
% ls =  2.250 (in)
```

The length of the thread that is in the clamp zone, l_t, is:

```
lt = lm - ls;
% lt =  0.750 (in)
```

The stiffness of the bolt is given by Eq. (3.29):

```
kb = (lt / (At*E) + ls / (As*E))^(-1);
% kb = 1.792e+06 (lb/in)
```

The stiffness of the steel clamped part with the same E is calculated with Eq. (3.40):

```
Asteel = 0.78715;
Bsteel = 0.62873;
km = E*d*Asteel*exp(Bsteel*d/lm);
% km = 1.311e+07 (lb/in)
```

The joint stiffness constant is a dimensionless stiffness parameter given by

```
C = kb / (km + kb);
% C = 0.120
```

The bolt preload (initial tensile force) is given by $F_i = KA_t S_p$, or in MAT-LAB as a function of K is [9]:

```
Fi = @(K) K * Sp * At;
```

where $K = 0.75$ for reused connections and $K = 0.90$ for permanent connections. The initial loads for reused and permanent connections are calculated with the MATLAB function Fi:

```
Fir = Fi(0.75);
% Fi reused = 6917.552 (lb)
Fip = Fi(0.9);
% Fi permanent = 8301.062 (lb)
```

The bolt axial load is given by the MATLAB function Fb:

```
Fb = @(K) Fi(K) + C * Fe;
```

The bolt axial loads for reused and permanent connections are:

```
Fbr = Fb(0.75);
% Fb reused = 7218.093 (lb)
Fbp = Fb(0.9);
% Fb permanent = 8601.603 (lb)
```

The clamping force is given by the MATLAB function Fc:

```
Fc = @(K) Fi(K) - (1 - C) * Fe;
```

and for reused and permanent connections is:

```
Fcr = Fc(0.75);
% Fc reused = 4718.093 (lb)
Fcp = Fc(0.9);
% Fc permanent = 6101.603 (lb)
```

The bolt factor against load is given by Eq. (3.50) and using a MATLAB function is:

```
nb = @(K) (At*Sp-Fi(K))/(C*Fe);
nbr = nb(0.75);
nbp = nb(0.90);
```

The numerical values for the bolt safety factor, n_b, are:

```
% nb reused    =   7.672
% nb permanent =   3.069
```

The load required to separate the joint is calculated in MATLAB as

```
Fs = @(K) Fi(K) / (1 - C);
Fsr = Fs(0.75);
Fsp = Fs(0.90);
% Fs reused    = 7862.788 (lb)
% Fs permanent = 9435.346 (lb)
```

and the safety factor against separation of the parts of the joint, n_s, is obtained from Eq. (3.51):

```
ns = @(K) Fs(K)/ Fe;
nsr = ns(0.75);
nsp = ns(0.90);
% ns reused    =   3.145
% ns permanent =   3.774
```

The factor for the bolt against the load, $n_b(K)$, and the safety factor against separation of the parts, $n_s(K)$, are defined as functions of the preload factor K [9]. The two factors are plotted in Fig. 3.19. The intersection of the two functions is obtained in MATLAB with:

```
syms x
xs = solve(nb(x)==ns(x));
Ko = eval(xs);
no = nb(Ko);
```

and the numerical values are:

```
% Ko =   0.880
% no =   3.689
```

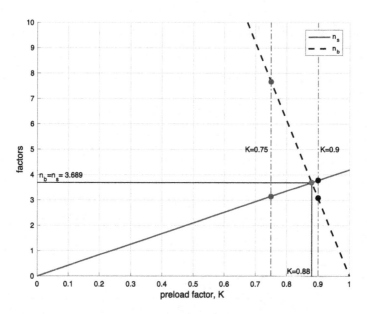

Figure 3.19 Load factor for the bolt, n_b, and safety factor against separation, n_s, in Example 3.4.

The load factor for the bolt, $n_b(K)$, increases linearly with the decrease of the preload factor K. The safety factor against separation of the parts, $n_s(K)$, increases linearly with the increase of the preload factor K. For a preload factor $K_0 = 0.880$ both factors are equal $n_0 = n_b = n_s = 3.689$. The MATLAB program for the graph in Fig. 3.19 is:

```
K = 0:0.05:1;
hold on
plot(K, ns(K),'-b ','LineWidth',2)
plot(K, nb(K),'--k','LineWidth',2)
xlabel('preload factor, K')
ylabel('factors')
ylim([0 10])
legend('n_s','n_b')
plot([0,Ko],[no,no],'-k','LineWidth',1)
plot([Ko,Ko],[0,no],'-k','LineWidth',1)
plot([.75,.75],[0,10],'-.b','LineWidth',1)
plot([.9,.9],[0,10],'-.r','LineWidth',1)
grid
plot(0.9,nsp,'ko',...
```

```
'MarkerSize',6,'LineWidth',2,'MarkerFaceColor','k')
plot(0.9,nbp,'ko',...
'MarkerSize',6,'LineWidth',2,'MarkerFaceColor','k')
plot(0.75,nsr,'bo',...
'MarkerSize',6,'LineWidth',2,'MarkerFaceColor','b')
plot(0.75,nbr,'bo',...
'MarkerSize',6,'LineWidth',2,'MarkerFaceColor','b')
plot(Ko,no,'ro',...
'MarkerSize',6,'LineWidth',2,'MarkerFaceColor','r')
text(.66,5,' K=0.75')
text(.9,5,' K=0.9')
text(0.795,.2,' K=0.88')
text(0,no+.25,' n_b=n_s= 3.689')
```

References

[1] E.A. Avallone, T. Baumeister, A. Sadegh, Marks' Standard Handbook for Mechanical Engineers, 11th edition, McGraw-Hill Education, New York, 2007.

[2] A. Bedford, W. Fowler, Dynamics, Addison Wesley, Menlo Park, CA, 1999.

[3] A. Bedford, W. Fowler, Statics, Addison Wesley, Menlo Park, CA, 1999.

[4] R. Budynas, K.J. Nisbett, Shigley's Mechanical Engineering Design, 9th edition, McGraw-Hill, New York, 2013.

[5] M. Choudhury, Member Stiffness of Bolted Joints, MSc Thesis, Georgia Institute of Technology, 1988.

[6] J.A. Collins, H.R. Busby, G.H. Staab, Mechanical Design of Machine Elements and Machines, 2nd edition, John Wiley & Sons, 2009.

[7] A. Ertas, J.C. Jones, The Engineering Design Process, John Wiley & Sons, New York, 1996.

[8] A.S. Hall, A.R. Holowenko, H.G. Laughlin, Schaum's Outline of Machine Design, McGraw-Hill, New York, 2013.

[9] B.G. Hamrock, B. Jacobson, S.R. Schmid, Fundamentals of Machine Elements, McGraw-Hill, New York, 1999.

[10] R.C. Hibbeler, Engineering Mechanics Statics & Dynamics, Pearson Education, Inc., Upper Saddle River, NJ, 2010.

[11] R.C. Juvinall, K.M. Marshek, Fundamentals of Machine Component Design, 5th edition, John Wiley & Sons, New York, 2010.

[12] K. Lingaiah, Machine Design Databook, 2nd edition, McGraw-Hill Education, New York, 2003.

[13] D.B. Marghitu, Mechanical Engineer's Handbook, Academic Press, San Diego, CA, 2001.

[14] D.B. Marghitu, M.J. Crocker, Analytical Elements of Mechanisms, Cambridge University Press, Cambridge, 2001.

[15] D.B. Marghitu, Kinematic Chains and Machine Component Design, Elsevier, Amsterdam, 2005.

[16] D.B. Marghitu, M. Dupac, N.H. Madsen, Statics with MATLAB, Springer, New York, NY, 2013.

[17] D.B. Marghitu, M. Dupac, Advanced Dynamics: Analytical and Numerical Calculations with MATLAB, Springer, New York, NY, 2012.

[18] D.B. Marghitu, Mechanisms and Robots Analysis with MATLAB, Springer, New York, NY, 2009.

[19] C.R. Mischke, Prediction of Stochastic Endurance Strength, Transaction of ASME, Journal Vibration, Acoustics, Stress, and Reliability in Design 109 (1) (1987) 113–122.

[20] R.L. Mott, Machine Elements in Mechanical Design, Prentice-Hall, Upper Saddle River, NJ, 1999.

[21] W.A. Nash, Strength of Materials, Schaum's Outline Series, McGraw-Hill, New York, 1972.

[22] R.L. Norton, Machine Design, Prentice-Hall, Upper Saddle River, NJ, 1996.

[23] R.L. Norton, Design of Machinery, McGraw-Hill, New York, 1999.

[24] W.C. Orthwein, Machine Component Design, West Publishing Company, St. Paul, 1990.

[25] D. Planchard, M. Planchard, SolidWorks 2013 Tutorial with Video Instruction, SDC Publications, 2013.

[26] C.A. Rubin, The Student Edition of Working Model, Addison-Wesley Publishing Company, Reading, MA, 1995.

[27] A.S. Seireg, S. Dandage, Empirical Design Procedure for the Thermodynamic Behavior of Journal Bearings, Journal of Lubrication Technology 104 (1982) 135–148.

[28] I.H. Shames, Engineering Mechanics – Statics and Dynamics, Prentice-Hall, Upper Saddle River, NJ, 1997.

[29] J.E. Shigley, C.R. Mischke, Mechanical Engineering Design, McGraw-Hill, New York, 1989.

[30] J.E. Shigley, C.R. Mischke, R.G. Budynas, Mechanical Engineering Design, 7th edition, McGraw-Hill, New York, 2004.

[31] J.E. Shigley, J.J. Uicker, Theory of Machines and Mechanisms, McGraw-Hill, New York, 1995.

[32] A.C. Ugural, Mechanical Design, McGraw-Hill, New York, 2004.

[33] J. Wileman, M. Choudhury, I. Green, Computation of Member Stiffness in Bolted Connections, Journal of Machine Design 193 (1991) 432–437.

[34] S. Wolfram, Mathematica, Wolfram Media/Cambridge University Press, Cambridge, 1999.

[35] Fundamentals of Engineering. Supplied-Reference Handbook, National Council of Examiners for Engineering and Surveying (NCEES), Clemson, SC, 2001.

[36] MatWeb – material property data, http://www.matweb.com/.

CHAPTER FOUR

Rolling bearings

4.1. Introduction

A bearing allows relative motion of joined elements. One of the elements can be fixed. Bearings can act as support for shaft and withstand radial and/or axial loads. The rolling elements, balls, rollers, or needles divide the bearing rings. The rolling bearings operate at lower starting friction with a coefficient of friction of $\mu = 0.001-0.003$. Other advantages are easy lubrication and replacement in case of failure. Some disadvantages of rolling bearings are collapse to large loads, higher costs, and noise.

Fig. 4.1 depicts the parts of rolling bearings: outer ring, inner ring, rolling element, and separator (retainer). The retainer keeps the distance between the rolling elements. The rings of the rolling bearing form the races of the bearing. A rolling element can have ball bearings (Fig. 4.1A), roller

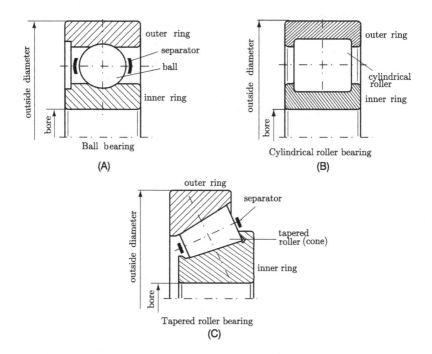

Figure 4.1 Parts of rolling bearings.

Machine component analysis with MATLAB®
https://doi.org/10.1016/B978-0-12-804229-8.00009-8

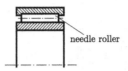

Figure 4.2 Needle bearings.

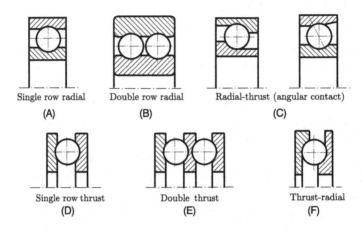

Figure 4.3 Ball bearings.

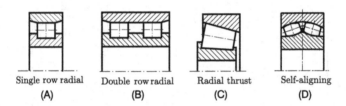

Figure 4.4 Roller bearings.

bearings (cylinder) (Fig. 4.1B), or tapered roller bearing (cone) (Fig. 4.1C). Needle bearings are shown in Fig. 4.2.

Different ball bearings are shown in Fig. 4.3, namely radial ball bearings (Figs. 4.3A–B), radial-thrust ball bearings (Fig. 4.3C), thrust ball bearings (Figs. 4.3D–E), and thrust-radial ball bearings (Fig. 4.3F). Fig. 4.4 shows different roller bearings.

Radial bearings sustain radial or transverse loads that are perpendicular to the axis of the shaft. Thrust bearings sustain axial or thrust loads that are parallel to the axis of the shaft. Single row ball bearings are designed mainly for transverse loads but can take some thrust loads. Radial-thrust or angular contact bearings are designed for significant axial loads. For notable loads, double row bearings are employed. For large deformations and tolerances of the shaft, self-aligning bearings are used.

Cylinder roller bearings have a larger area of contact relative to ball bearings and are designed for greater loads. Cylinder roller bearings are not designed for axial loads. Tapered or cone roller bearings can sustain transverse or axial forces. Needle bearings are designed for restricted areas and can be without separators.

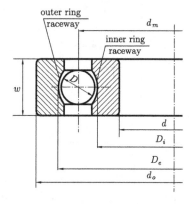

Figure 4.5 Geometry of a ball bearing.

The principal geometry elements of a ball bearing are shown in Fig. 4.5 and are defined as follows. The pitch diameter is calculated with

$$d_m \approx \frac{d_0 + d}{2}, \tag{4.1}$$

where d_0 is the outer diameter of the ball bearing and d is the bore. Using the diameter of the race of the outer ring, D_e, and the diameter of the race of the inner ring, D_i, the pitch diameter is given by the following relation:

$$d_m = \frac{D_i + D_e}{2}. \tag{4.2}$$

The diametral clearance between the balls of diameter D and the raceways is given by

$$s_d = D_e - D_i - 2D. \tag{4.3}$$

4.2. Force analysis

An angular contact or radial thrust ball bearing with the contact angle α is depicted in Fig. 4.6. The contact angle for a radial ball bearing is null.

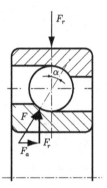

Figure 4.6 Radial thrust (angular contact) ball bearing.

A transverse load, F_r, is acting perpendicular to the shaft axis of the angular contact bearing. The force that acts on the ball is

$$F = \frac{F_r}{\cos \alpha}. \tag{4.4}$$

The thrust or axial load is

$$F_a = F_t = F \sin \alpha. \tag{4.5}$$

4.3. Catalogue bearing dimensions

Fig. 4.5 shows some standard dimensions of the ball bearings [9]: the bore d, the outside diameter d_0, and the width w. There are different series for bearings: extra-extra-light or LL00, extra-light or L00, light or 200, and medium or 300 [9]. The bearing basic number BBN for some ball bearing in terms of d and d_0 are given by [9]

$$BBN = \begin{cases}
\text{L02} & \text{for } d = 15 \text{ mm and } d_0 = 32 \text{ mm,} \\
\text{202} & \text{for } d = 15 \text{ mm and } d_0 = 35 \text{ mm,} \\
\text{302} & \text{for } d = 15 \text{ mm and } d_0 = 42 \text{ mm,} \\[6pt]
\text{L04} & \text{for } d = 20 \text{ mm and } d_0 = 42 \text{ mm,} \\
\text{204} & \text{for } d = 20 \text{ mm and } d_0 = 47 \text{ mm,} \\
\text{304} & \text{for } d = 20 \text{ mm and } d_0 = 52 \text{ mm,} \\[6pt]
\text{L05} & \text{for } d = 25 \text{ mm and } d_0 = 47 \text{ mm,} \\
\text{205} & \text{for } d = 25 \text{ mm and } d_0 = 52 \text{ mm,} \\
\text{305} & \text{for } d = 25 \text{ mm and } d_0 = 62 \text{ mm,} \\[6pt]
\text{L06} & \text{for } d = 30 \text{ mm and } d_0 = 55 \text{ mm,} \\
\text{206} & \text{for } d = 30 \text{ mm and } d_0 = 62 \text{ mm,} \\
\text{306} & \text{for } d = 30 \text{ mm and } d_0 = 72 \text{ mm,} \\[6pt]
\text{L07} & \text{for } d = 35 \text{ mm and } d_0 = 62 \text{ mm,} \\
\text{207} & \text{for } d = 35 \text{ mm and } d_0 = 72 \text{ mm,} \\
\text{307} & \text{for } d = 35 \text{ mm and } d_0 = 80 \text{ mm,} \\[6pt]
\text{L08} & \text{for } d = 40 \text{ mm and } d_0 = 68 \text{ mm,} \\
\text{208} & \text{for } d = 40 \text{ mm and } d_0 = 80 \text{ mm,} \\
\text{308} & \text{for } d = 40 \text{ mm and } d_0 = 90 \text{ mm,} \\[6pt]
\text{L09} & \text{for } d = 45 \text{ mm and } d_0 = 75 \text{ mm,} \\
\text{209} & \text{for } d = 45 \text{ mm and } d_0 = 85 \text{ mm,} \\
\text{309} & \text{for } d = 45 \text{ mm and } d_0 = 100 \text{ mm,} \\[6pt]
\text{L10} & \text{for } d = 50 \text{ mm and } d_0 = 80 \text{ mm,} \\
\text{210} & \text{for } d = 50 \text{ mm and } d_0 = 90 \text{ mm,} \\
\text{310} & \text{for } d = 50 \text{ mm and } d_0 = 110 \text{ mm,} \\[6pt]
\text{L17} & \text{for } d = 85 \text{ mm and } d_0 = 130 \text{ mm,} \\
\text{217} & \text{for } d = 85 \text{ mm and } d_0 = 150 \text{ mm,} \\
\text{317} & \text{for } d = 85 \text{ mm and } d_0 = 180 \text{ mm.}
\end{cases} \tag{4.6}$$

4.4. Catalogue selection for bearings

The geometry of the bearings and the standard forces are found in catalogues where the bearings are tabulated. Next a procedure to select the rolling bearing is presented [9].

The life of a rolling bearing represents the number of revolutions or the period of time at a constant angular velocity prior to the initial sign of material failure. Table 4.1 shows the rated capacity, C, corresponding to a standard life of $L_R = 9\,(10^7)$ revolutions, and a constant radial load.

Bearing application required life is calculated with

$$L = L_R \left(\frac{C}{F_r}\right)^{10/3}, \tag{4.7}$$

where the radial load for the application is F_r. The required rated capacity for the application is

$$C_{req} = F_r \left(\frac{L}{L_R}\right)^{3/10}. \tag{4.8}$$

The standard life denoted by L_{10} or B_{10} is the life of the bearing that corresponds to a reliability of $r = 90\%$ or 10% failure. The median life is five times the standard life. Eqs. (4.7) and (4.8) for ball and roller bearings are improved with a reliability factor K_r [9]

$$L = K_r L_R \,(C/F_r)^{3.33},$$
$$C_{req} = F_r \left(\frac{L}{K_r L_R}\right)^{0.3}, \tag{4.9}$$

where

$$K_r = \begin{cases} 1.00 & \text{for 90\% reliability, } L_{10}, \\ 0.62 & \text{for 95\% reliability, } L_5, \\ 0.53 & \text{for 96\% reliability, } L_4, \\ 0.44 & \text{for 97\% reliability, } L_3, \\ 0.33 & \text{for 98\% reliability, } L_2, \\ 0.21 & \text{for 99\% reliability, } L_1. \end{cases} \tag{4.10}$$

Table 4.1 Bearing rated capacities, C, for $L_R = 9\,(10^7)$ revolutions with 90% reliability. From New Departure-Hyatt Bearing Division, General Motors Corporation. Used with permission from General Motors Corporation Inc.

	Radial ball, $\alpha = 0°$			Angular ball, $\alpha = 25°$			Roller		
	L00	200	300	L00	200	300	1000	1200	1300
d	xlt	lt	med	xlt	lt	med	xlt	lt	med
mm	kN	kN	kN	kN	kN	kN	kN	kN	kN
10	1.02	1.42	1.90	1.02	1.10	1.88			
12	1.12	1.42	2.46	1.10	1.54	2.05			
15	1.22	1.56	3.05	1.28	1.66	2.85			
17	1.32	2.70	3.75	1.36	2.20	3.55	2.12	3.80	4.90
20	2.25	3.35	5.30	2.20	3.05	5.80	3.30	4.40	6.20
25	2.45	3.65	5.90	2.65	3.25	7.20	3.70	5.50	8.50
30	3.35	5.40	8.80	3.60	6.00	8.80		8.30	10.0
35	4.20	8.50	10.6	4.75	8.20	11.0		9.30	13.1
40	4.50	9.40	12.6	4.95	9.90	13.2	7.20	11.1	16.5
45	5.80	9.10	14.8	6.30	10.4	16.4	7.40	12.2	20.9
50	6.10	9.70	15.8	6.60	11.0	19.2		12.5	24.5
55	8.20	12.0	18.0	9.00	13.6	21.5	11.3	14.9	27.1
60	8.70	13.6	20.0	9.70	16.4	24.0	12.0	18.9	32.5
65	9.10	16.0	22.0	10.2	19.2	26.5	12.2	21.1	38.3
70	11.6	17.0	24.5	13.4	19.2	29.5		23.6	44.0
75	12.2	17.0	25.5	13.8	20.0	32.5		23.6	45.4
80	14.2	18.4	28.0	16.6	22.5	35.5	17.3	26.2	51.6
85	15.0	22.5	30.0	17.2	26.5	38.5	18.0	30.7	55.2
90	17.2	25.0	32.5	20.0	28.0	41.5		37.4	65.8
95	18.0	27.5	38.0	21.0	31.0	45.5		44.0	65.8
100	18.0	30.5	40.5	21.5	34.5		20.9	48.0	72.9
105	21.0	32.0	43.5	24.5	37.5			49.8	84.5
110	23.5	35.0	46.0	27.5	41.0	55.0	29.4	54.3	85.4
120	24.5	37.5		28.5	44.5			61.4	100.1
130	29.5	41.0		33.5	48.0	71.0	48.9	69.4	120.1
140	30.5	47.5		35.0	56.0			77.4	131.2
150	34.5			39.0	62.0		58.7	83.6	
160								113.4	
180	47.0			54.0			97.9	140.1	
200								162.4	
220								211.3	
240								258.0	

The influence of the axial force is adjusted with an radial equivalent force, F_e [9]

$$L = K_r L_R \left(\frac{C}{F_e}\right)^{3.33},$$

$$C_{req} = F_e \left(\frac{L}{K_r L_R}\right)^{0.3}. \qquad (4.11)$$

The radial equivalent load for ball bearings with the contact angle $\alpha = 0°$ is

$$F_e = \begin{cases} F_r & \text{for } 0.00 < F_a/F_r < 0.35, \\ F_r \left[1 + 1.115 \left(\dfrac{F_a}{F_r} - 0.35\right)\right] & \text{for } 0.35 < F_a/F_r < 10.0, \\ 1.176\, F_a & \text{for } F_a/F_r > 10.0. \end{cases} \qquad (4.12)$$

For angular ball bearings with the contact angle $\alpha = 25°$, the radial equivalent load is

$$F_e = \begin{cases} F_r & \text{for } 0.00 < F_a/F_r < 0.68, \\ F_r \left[1 + 0.87 \left(\dfrac{F_a}{F_r} - 0.68\right)\right] & \text{for } 0.68 < F_a/F_r < 10.0, \\ 0.911\, F_a & \text{for } F_a/F_r > 10.0. \end{cases} \qquad (4.13)$$

The influence of a shock force is taken into consideration by an application factor K_a [4,9]:

$$L = K_r L_R \left(\frac{C}{K_a F_e}\right)^{3.33},$$

$$C_{req} = K_a F_e \left(\frac{L}{K_r L_R}\right)^{0.3}. \qquad (4.14)$$

The application factor for a ball bearing is

$$K_a = \begin{cases} 1 & \text{for uniform force, no impact,} \\ 1.0\text{--}1.1 & \text{for precision gearing,} \\ 1.1\text{--}1.3 & \text{for commercial gearing,} \\ 1.2 & \text{for poor bearing seals,} \\ 1.2\text{--}1.5 & \text{for light impact,} \\ 1.5\text{--}2.0 & \text{for moderate impact,} \\ 2.0\text{--}3.0 & \text{for heavy impact.} \end{cases} \qquad (4.15)$$

The application factor for a roller bearing is

$$K_a = \begin{cases} 1 & \text{for uniform force, no impact,} \\ 1 & \text{for gearing,} \\ 1.0\text{--}1.1 & \text{for light impact,} \\ 1.1\text{--}1.5 & \text{for moderate impact,} \\ 1.5\text{--}2.0 & \text{for heavy impact.} \end{cases} \tag{4.16}$$

The following suggestions for the life of the bearing, in kh, are given [4,9]:

$$\text{design life} = \begin{cases} 0.1\text{--}0.5 \text{ kh} & \text{for instruments and apparatus} \\ & \text{for infrequent use,} \\ 0.5\text{--}2.0 \text{ kh} & \text{for aircraft engines,} \\ 4\text{--}8 \text{ kh} & \text{for machines used intermittently,} \\ & \text{where service interruption} \\ & \text{is of minor importance,} \\ 8\text{--}14 \text{ kh} & \text{for machines used intermittently,} \\ & \text{where reliability} \\ & \text{is of great importance,} \\ 14\text{--}20 \text{ kh} & \text{for machines for 8-hour service,} \\ & \text{but not every day,} \\ 20\text{--}30 \text{ kh} & \text{for machines for 8-hour service,} \\ & \text{every working day,} \\ 50\text{--}60 \text{ kh} & \text{for machines for continuous} \\ & \text{24-hour service,} \\ 100\text{--}200 \text{ kh} & \text{for machines for continuous} \\ & \text{24-hour service where} \\ & \text{reliability is of extreme importance.} \end{cases} \tag{4.17}$$

4.5. Examples

Example 4.1. A gear force $F = 1\,000$ lb acts on a shaft that rotates at $n = 300$ rpm, as shown in Fig. 4.7. The dimensions of the shaft are $a = 1.25$ ft and $b = 0.25$ ft. The machine is intended for 8-hour service, every working day. A reliability of 90% is desired. Select identical 300 series ball bearings for A and B.

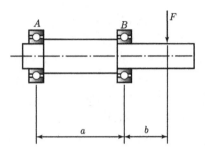

Figure 4.7 Example 4.1.

Solution

The radial loads on the shaft at A and B need to be calculated. The radial load at bearing B, F_B, is calculated using the sum of the moments with respect to A:

```
F = 1000; % (lb)
a = 1.25; % (ft)
b = 0.25; % (ft)
FB = F*(a+b)/a;
% FB = 1200.000 (lb)
```

The radial force at bearing A, F_A, is calculated from the sum of the forces for the shaft:

```
FA = -F+FB;
% FA = 200.000 (lb)
```

The radial load on the bearing at B is greater than the radial load on the bearing at A the selection of the bearings will be made on bearing B, $F_r = F_B$:

```
if FB > FA
Fr = FB;
else
Fr = FA;
end
```

The radial equivalent force is $F_e = F_r = F_B = 1200.000$ (lb). The application factor is given by Eq. (4.15) and for gearing it is selected as $K_a = 1.1$. The life adjustment reliability factor for 90% reliability with Eq. (4.10) is $K_r = 1$.

The life of a bearing for machines for 8-hour service every working day, with Eq. (4.17) is $L_d = 30$ kh. The corresponding life of bearings for 30000 h operation at a speed of 300 rpm is:

```
Ld = 30000; % (h)
n = 300;    % (rpm)
L = Ld * n * 60;
% L = 5.400e+08 (rev)
```

The required value of the rated capacity for the application is:

```
LR = 90*10^6;
Creq = Fe*ka*(L/(kr*LR))^0.3;
% Creq = 2.260e+03 (lb)
% 1 N = 0.224808942443 lb
Creq = Creq/0.224808942443; % (N)
% Creq = 1.005e+04 (N)
```

From Table 4.1 with $C = 10.05$ kN for 300 series select $d = 35$ mm. From Eq. (4.6) with $d = 35$ mm for 300 series select a bearing number 307.

Example 4.2. Fig. 4.8 shows a gear with the pitch radius $R = OP = 0.10$ m. The uniform forces acting on the gear at P are $F_{Px} = F_x = 1500$ N, $F_{Py} = F_y = 1700$ N, and $F_{Pz} = F_z = 5000$ N. The directions of the forces are represented in Fig. 4.8. The bearing at A can take axial (thrust) forces. The shaft rotates at a speed of $n = 1\,050$ rpm and has the dimensions $a = 0.15$ m and $b = 0.07$ m. The shaft external diameter is 30 mm at A and B. The design life is to be $DL = 5\,000$ h and an application factor of 1 is suitable. The reliability is 90%. Select one bearing size suitable for both locations A and B.

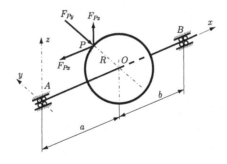

Figure 4.8 Example 4.2.

Solution

The input data in MATLAB® are:

```
Fx = -1500; % (N)
Fy = -1700; % (N)
Fz =  5000; % (N)
FP_= [Fx,Fy,Fz]; % gear force vector at P
a = 0.15;   % (m)
b = 0.07;   % (m)
R = 0.10;   % (m)
rA_ = [0, 0, 0]; % position vector of bearing A
rB_ = [a+b, 0,0]; % position vector of bearing B
rO_ = [a, 0,0];  % position vector of gear center
rP_ = [a, R, 0]; % position vector of gear forces
```

The unknowns are the bearing reactions at *A* and *B*

```
syms FAx FAy FAz FBx FBy FBz
FA_ = [FAx, FAy, FAz]; % reaction force at A
FB_ = [  0, FBy, FBz]; % reaction force at B
```

The reaction force at *B* is calculated using the sum of the moments with respect to *A*:

```
% sum MA_ = rP_ x FP_ + rB_ x FB_  = 0_ =>
MA_ = cross(rP_, FP_) + cross(rB_,FB_);
solFB = solve(MA_(2),MA_(3));
FBys=eval(solFB.FBy);
FBzs=eval(solFB.FBz);
FB_ = [0, FBys, FBzs];
% FB_ = [ 0 477.273 -3409.091] (N)
```

The radial force at *B* is:

```
FBr = sqrt(FB_(2)^2+FB_(3)^2);
% radial force FBr = 3442.338 (N)
```

The reaction bearing force at *A* is calculated using the sum of the forces for the system:

```
% sum F_ = FA_+FB_+FP_ = 0_ =>
FA_ = -FP_-FB_;
% FA_ = [ 1500.000 1222.727 -1590.909] (N)
```

and the radial and axial forces on the bearing at A are:

```
FAr = sqrt(FA_(2)^2+FA_(3)^2);
FAa = FA_(1);
% radial force FAr = 2006.503 (N)
% axial force FAa = 1500.000 (N)
```

The MATLAB program for the graph of the free body diagram of the system is:

```
% scale factor for forces
sf = 20000;
af = 1;
axis([-af af -af af -af af])
grid on
hold on
axis auto
xlabel('x(m)'), ylabel('y(m)'), zlabel('z(m)')
t = linspace(0,2*pi);
% plot gear of radius R at 0
plot3(r0_(1)+0*t,r0_(2)+R*cos(t),r0_(3)+R*sin(t),...
 'color','k','LineWidth',3)
text(rA_(1),rA_(2),rA_(3),'   A','fontsize',12)
text(rB_(1),rB_(2),rB_(3),'   B','fontsize',12)
text(rP_(1),rP_(2),rP_(3),'   P','fontsize',12)
text(r0_(1),r0_(2),r0_(3),'   0','fontsize',12)
line([rA_(1),rB_(1)],[rA_(2),rB_(2)],[rA_(3),rB_(3)],...
    'LineStyle','-','LineWidth',2)
line([r0_(1),rP_(1)],[r0_(2),rP_(2)],[r0_(3),rP_(3)],...
    'LineStyle','-','LineWidth',1)
% plot FP_ at P
quiver3(...
 rP_(1),rP_(2),rP_(3),...
 FP_(1)/sf,0,0,'color','b','LineWidth',1.3);
quiver3(...
 rP_(1),rP_(2),rP_(3),...
 0,FP_(2)/sf,0,'color','b','LineWidth',1.3);
quiver3(...
 rP_(1),rP_(2),rP_(3),...
  0,0,FP_(3)/sf,'color','b','LineWidth',1.3);
```

```
text(rP_(1)+FP_(1)/sf,rP_(2),rP_(3),'F_x','fontsize',12)
text(rP_(1),rP_(2)+FP_(2)/sf+0.025,rP_(3),'F_y','fontsize',12)
text(rP_(1),rP_(2),rP_(3)+FP_(3)/sf,'F_z','fontsize',12)
% plot FA_ at A
quiver3(...
 rA_(1),rA_(2),rA_(3),...
 FA_(1)/sf,0,0,'color','r','LineWidth',1.3);
quiver3(...
 rA_(1),rA_(2),rA_(3),...
 0,FA_(2)/sf,0,'color','r','LineWidth',1.3);
quiver3(...
 rA_(1),rA_(2),rA_(3),...
  0,0,FA_(3)/sf,'color','r','LineWidth',1.3);
text(rA_(1)+FA_(1)/sf,rA_(2),rA_(3),'F_{Ax}','fontsize',12)
text(rA_(1),rA_(2)+FA_(2)/sf,rA_(3),'F_{Ay}','fontsize',12)
text(rA_(1),rA_(2),rA_(3)+FA_(3)/sf,'F_{Az}','fontsize',12)
% plot FB_ at B
quiver3(...
 rB_(1),rB_(2),rB_(3),...
 0,FB_(2)/sf,0,'color','r','LineWidth',1.3);
quiver3(...
 rB_(1),rB_(2),rB_(3),...
  0,0,FB_(3)/sf,'color','r','LineWidth',1.3);
text(rB_(1),rB_(2)+FB_(2)/sf,rB_(3),'F_{By}','fontsize',12)
text(rB_(1),rB_(2),rB_(3)+FB_(3)/sf,'F_{Bz}','fontsize',12)
plot3(r0_(1),r0_(2),r0_(3),'o','color','k')
plot3(rA_(1),rA_(2),rA_(3),'o','color','k')
plot3(rB_(1),rB_(2),rB_(3),'o','color','k')
view(-20,16)
```

The free body diagram using MATLAB is shown in Fig. 4.9. The bearing with a shaft speed of $n = 1050$ rpm will last for

```
DL = 5000; % (h) design life
n  = 1050; % (rpm)
L = n * DL * 60;
% life L = 3.15e+08 (rev)
```

With 90% reliability, the reliability factor is $K_r = 1$. The life corresponding to rated capacity is $L_R = 90 (10)^6$. The application factor for uniform force is $K_a = 1$. The bearing inner bore size is $d = 30$ mm.

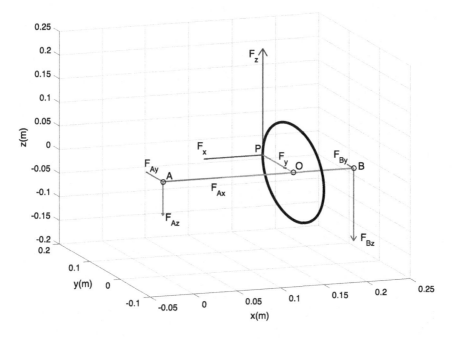

Figure 4.9 Example 4.2 free body diagram with MATLAB.

Analysis for bearing at A

For ball bearings, the radial equivalent force, F_e, is calculated using Eq. (4.12):

```
fr=FAa/FAr;
% FAa/FAr =  0.748
FAe = FAr*(1+1.115*(FAa/FAr-0.35));
% radial equivalent force FAe = 2895.965 (kN)
```

For radial ball bearings, the required value of rated capacity for the application is:

```
CreqA = Ka*FAe*(L/(Kr*LR))^0.3*10^-3;
% rated capacity: CreqA =  4.217 (kN)
```

From Table 4.1, with a rated capacity of 5.40 kN, a 206 radial ball bearing will be acceptable for A.

The radial equivalent force, F_e, for angular ball bearings with $\alpha = 25°$ is given by Eq. (4.13):

```
% FAa/FAr =  0.748
FAae = FAr*(1+0.87*(FAa/FAr-0.68));
% angular ball bearing: FAae = 2124.456 (kN)
CreqAa = Ka*FAae*(L/(Kr*LR))^0.3*10^-3;
% angular ball bearing: CreqAa =  3.094 (kN)
```

From Table 4.1, with a rated capacity of 6 kN, a 206 angular ball bearing will be acceptable for A.

Analysis for bearing at B
For the radial ball bearing at B, the radial equivalent force is:

```
FBe = FBr;
% radial equivalent force FBe = 3442.338 (kN)
```

and the rated capacity is calculated with:

```
CreqB = Ka*FBe*(L/(Kr*LR))^0.3*10^-3;
% rated capacity: CreqB =  5.013 (kN)
```

From Table 4.1, with a rated capacity of 5.40 kN, a 206 radial ball bearing will be acceptable for B.

The capacity of a 206 radial ball bearing with a 30 mm inner diameter is suitable at both position A and B.

Example 4.3. Fig. 4.10 depicts a countershaft with two rigidly connected gears 1 and 2. The angular speed of the countershaft is 200 rpm. The force on the countershaft gear 1 at P is $\mathbf{F}_P = F_{Py}\mathbf{J} + F_{Pz}\mathbf{k} = 1500\mathbf{J} + 1500\tan(20°)\mathbf{k}$ N, and the force on the gear 2 at R is $\mathbf{F}_R = F_{Ry}\mathbf{J} + F_{Rz}\mathbf{k} = -2000\tan(20°)\mathbf{J} - 2000\mathbf{k}$ N. The pitch radius of gear 1 is $OP = R_1 = 0.20$ m and the pitch radius of gear 2 is $CR = R_2 = 0.15$ m. The distance between the bearings is $s = 0.20$ m and the distance between the gear and bearing is $l = 0.15$ m. The gear system is a part of an industrial machine intended for 8-hour service, but not every day. Select identical 300 series ball bearings for A and B.

Solution
The forces and their position vectors are given by:

```
FPx = 0;      % (N)
FPy = 1500;   % (N)
FPz = FPy*tan(20*pi/180);
```

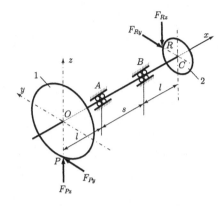

Figure 4.10 Example 4.3.

```
FP_= [FPx,FPy,FPz];
% external force FP_ = [0 1500.000 545.955] (N)
FRx = 0;      % (N)
FRz = -2000;  % (N)
FRy = FRz*tan(20*pi/180);
FR_= [FRx,FRy,FRz];
% external force FR_ = [0 -727.940 -2000.000] (N)
R1 = 0.20; % (m) pitch radius of gear 1
R2 = 0.15; % (m) pitch radius of gear 2
l = 0.15; % (m)
s = 0.20; % (m)
r0_ = [0, 0, 0]; % origin, center of gear 1
rA_ = [l, 0, 0]; % position of bearing at A
rB_ = [l+s, 0, 0]; % position of bearing at B
rC_ = [2*l+s, 0, 0]; % position of center of gear 2
rP_ = [0, 0, -R1]; % application point of force FP_
rR_ = [2*l+s, R2, 0]; % application point of force FR_
```

The unknowns are the bearing loads at A and B:

```
syms   FAy FAz FBy FBz
FA_ = [0, FAy, FAz];
FB_ = [0, FBy, FBz];
```

The bearing reaction force at B is calculated using the sum of the moments about A:

```
% sum MA_ = rAP_ x FP_ + rAB_ x FB_  + rAR_ x FR_ = 0_ =>
MA_ = cross(rP_-rA_, FP_) + cross(rB_-rA_,FB_) + cross(rR_-rA_,FR_);
digits(6)
MAx=vpa(MA_(1));
fprintf('MAx = %s \n',char(MAx))
MAy=vpa(MA_(2));
fprintf('MAy = %s \n',char(MAy))
MAz=vpa(MA_(3));
fprintf('MAz = %s \n',char(MAz))
solFB = solve(MA_(2),MA_(3));
FBys=eval(solFB.FBy);
FBzs=eval(solFB.FBz);
FB_ = [0, FBys, FBzs];
FBr = sqrt(FB_(2)^2+FB_(3)^2);
% sum M_Ay = 781.893 - 0.2*FBz
% sum M_Az = 0.2*FBy - 479.779
% FB_ = [ 0 2398.896 3909.467] (N)
% radial force FBr = 4586.789 (N)
```

The bearing reaction force at A is calculated using the sum of the forces for the system:

```
% sum F_ = FA_+FB_+FP_+FR_= 0_ =>
FA_ = -FP_-FB_-FR_;
FAr = sqrt(FA_(2)^2+FA_(3)^2);
% FA_ = [ -0.000 -3170.955 -2455.422] (N)
% radial force FAr = 4010.493 (N)
```

A representation of the free body diagram is shown in Fig. 4.11 and is obtained using the MATLAB commands:

```
sf = 8000; % scale factor for forces
af = .5;
axis([-af af -af af -af af])
grid on
hold on
axis auto
xlabel('x(m)'), ylabel('y(m)'), zlabel('z(m)')
t = linspace(0,2*pi);
% plot gear 1 of radius R1 at 0
plot3(0*t,R1*cos(t),R1*sin(t),...
  'color','k','LineWidth',3)
```

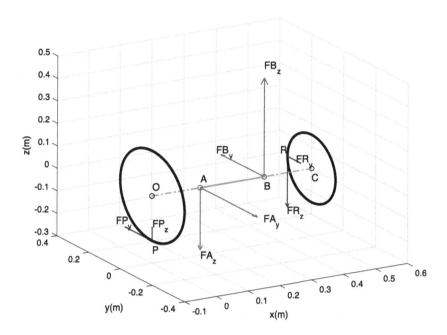

Figure 4.11 Free body diagram of Example 4.3 obtained with MATLAB.

```
% plot gear 2 of radius R2 at C
plot3(rC_(1)+0*t,rC_(2)+R2*cos(t),rC_(3)+R2*sin(t),...
 'color','k','LineWidth',3)
plot3(0,0,0,'o','color','k')
plot3(rC_(1),rC_(2),rC_(3),'o','color','k')
plot3(rA_(1),rA_(2),rA_(3),'o','color','b')
plot3(rB_(1),rB_(2),rB_(3),'o','color','b')
line([rO_(1),rC_(1)],[rO_(2),rC_(2)],[rO_(3),rC_(3)],...
    'LineStyle','-.','LineWidth',1.5)
line([rA_(1),rB_(1)],[rA_(2),rB_(2)],[rA_(3),rB_(3)],...
    'LineStyle','-','LineWidth',2)
% plot FP_ at P
quiver3(...
 rP_(1),rP_(2),rP_(3),...
 0,FP_(2)/sf,0,'color','b','LineWidth',1.3);
quiver3(...
 rP_(1),rP_(2),rP_(3),...
  0,0,FP_(3)/sf,'color','b','LineWidth',1.3);
% plot FR_ at R
```

```
quiver3(...
 rR_(1),rR_(2),rR_(3),...
 0,FR_(2)/sf,0,'color','b','LineWidth',1.3);
quiver3(...
 rR_(1),rR_(2),rR_(3),...
  0,0,FR_(3)/sf,'color','b','LineWidth',1.3);
% plot FA_ at A
quiver3(...
 rA_(1),rA_(2),rA_(3),...
 0,FA_(2)/sf,0,'color','r','LineWidth',1.3);
quiver3(...
 rA_(1),rA_(2),rA_(3),...
  0,0,FA_(3)/sf,'color','r','LineWidth',1.3);
% plot FB_ at B
quiver3(...
 rB_(1),rB_(2),rB_(3),...
 0,FB_(2)/sf,0,'color','r','LineWidth',1.3);
quiver3(...
 rB_(1),rB_(2),rB_(3),...
  0,0,FB_(3)/sf,'color','r','LineWidth',1.3);
text(rP_(1),rP_(2),rP_(3)-0.04,'P','fontsize',12)
text(rP_(1),rP_(2)+FP_(2)/sf+0.04,rP_(3),'FP_y','fontsize',12)
text(rP_(1),rP_(2),rP_(3)+FP_(3)/sf,'FP_z','fontsize',12)
text(rR_(1),rR_(2)+0.08,rR_(3),'  R','fontsize',12)
text(rR_(1),rR_(2)+FR_(2)/sf+0.04,rR_(3),'FR_y','fontsize',12)
text(rR_(1),rR_(2),rR_(3)+FR_(3)/sf,'FR_z','fontsize',12)
text(rA_(1),rA_(2),rA_(3)+0.04,'A','fontsize',12)
text(rA_(1),rA_(2)+FA_(2)/sf,rA_(3),'FA_y','fontsize',12)
text(rA_(1),rA_(2),rA_(3)+FA_(3)/sf,'FA_z','fontsize',12)
text(rB_(1),rB_(2),rB_(3)-0.04,'B','fontsize',12)
text(rB_(1),rB_(2)+FB_(2)/sf,rB_(3),'FB_y','fontsize',12)
text(rB_(1),rB_(2),rB_(3)+FB_(3)/sf,'FB_z','fontsize',12)
text(r0_(1),r0_(2),r0_(3)+0.04,'0','fontsize',12)
text(rC_(1),rC_(2),rC_(3)-0.04,'C','fontsize',12)
view(-30,23)
```

The radial force at bearing *B* is higher than that at bearing *A*, and the selection of the bearing is based on this force FBr = 4586.789 (N):

```
if FAr > FBr
        Fe = FAr;
else
        Fe = FBr;
end
% radial equivalent force Fe = FBr = 4586.789 (N)
```

From Eq. (4.17) the design bearing life for machines for 8-hour service, but not every day, is selected (conservatively) $DL = 20\,000$ h. The life L corresponding to the 200 rpm rotation of the shaft is:

```
DL = 20000; % (h) design life
n =    200; % (rpm)
L = n * DL * 60;
% life L = 2.400e+08 (rev)
```

The life adjustment is represented by the reliability factor K_r, given by Eq. (4.10), $K_r = 0.33$ for 98% reliability. The application factor for commercial gearing is selected (conservatively) from Eq. (4.15), $K_a = 1.3$. The life corresponding to rated capacity is $L_R = 90\,(10)^6$. The required value of rated capacity for this application is given by:

```
Ka = 1.3;  % for commercial gearing
Kr = 0.33; % r = 98%
LR = 90 * 10^6;
Creq = Ka*Fe*(L/(Kr*LR))^0.3*10^-3;
% rated capacity: Creq = 11.161 (kN)
```

From Table 4.1 with 11.16 kN for 300 series we select $C = 12.6$ kN and $d = 40$ mm bore. From Eq. (4.6) with 40 mm bore and 300 series the bearing number is 308.

A bearing number 308 is selected for both bearings A and B.

Example 4.4. Radial forces act on a number 317 ($d = 85$ 5mm) radial contact ball bearing as follows: 25% of the time $F_{r1} = 5$ kN at $n_{r1} = 1000$ rpm, 25% of the time $F_{r2} = 10$ kN at $n_{r2} = 1200$ rpm, 25% of the time $F_{r3} = 15$ kN at $n_{r3} = 1500$ rpm, and 25% of the time $F_{r4} = 20$ kN at $n_{r2} = 500$ rpm. The loads are with moderate impact and the reliability is $r = 96\%$. Estimate the bearing life.

Solution

For the number 317 radial contact bearing from Table 4.1, the rated capacity is $C = 30$ kN. The input data in MATLAB is:

```
Cap = 30; % (kN)
nr1 = 1000; % (rpm)
t1  = 0.25; % 25% time
Fr1 = 5;    % (kN)
nr2 = 1200; % (rpm)
t2  = 0.25; % 25% time
Fr2 = 10;   % (kN)
nr3 = 1500; % (rpm)
t3  = 0.25; % 25% time
Fr3 = 15;   % (kN)
nr4 = 500;  % (rpm)
t4  = 0.25; % 25% time
Fr4 = 20;   % (kN)
```

The radial load versus time is calculated with MATLAB:

```
x = 0 : 2.5 : 100;
for ix = 1 : length(x)
    y(ix) = piecewiseF(x(ix));
end
stem(x, y,'filled')
stem(x, y,...
    'LineStyle','-',...
    'LineWidth',1.5,...
    'MarkerFaceColor','b',...
    'MarkerEdgeColor','b')
axis([0 100 0 22])
ylabel('F_r (kN)','FontSize',12)
xlabel('%','FontSize',12)
ax = gca;
ax.XTick = [0,25,50,75,100];
ax.YTick = [5, 10, 15, 20];
set(gca, ...
'YTickLabel',num2str(get(gca,'YTick')', '%4.0f'));
set(gca, ...
'XTickLabel',num2str(get(gca,'XTick')', '%4.0f'));
grid
```

where the piecewise force function `piecewiseF` is

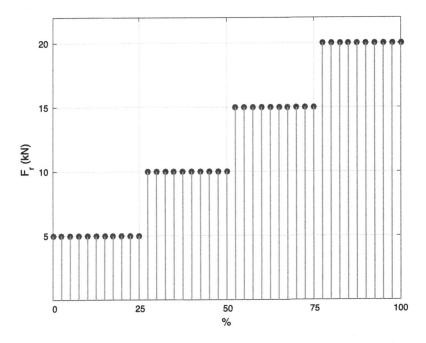

Figure 4.12 Example 4.4.

```
function y = piecewiseF(x)
if x <= 25
    y = 5;
elseif 25 < x & x <= 50
    y = 10;
elseif 50 < x & x <= 75
    y = 15;
else
    y = 20;
end
end
```

The MATLAB graph of the variable force is shown in Fig. 4.12. The application factor is $K_a = 2$ for moderate impact, and for 96% reliability the factor is $K_r = 0.53$. The corresponding life for each radial force is calculated with:

```
Ka = 2;    % moderate impact
Kr = 0.53; % 96% reliability
```

```
LR = 90*10^6;
L1 = Kr*LR*( Cap / (Ka*Fr1))^(10/3);
L2 = Kr*LR*( Cap / (Ka*Fr2))^(10/3);
L3 = Kr*LR*( Cap / (Ka*Fr3))^(10/3);
L4 = Kr*LR*( Cap / (Ka*Fr4))^(10/3);
% L1 = 1.857e+09 (rev)
% L2 = 1.843e+08 (rev)
% L3 = 4.770e+07 (rev)
% L4 = 1.828e+07 (rev)
```

Let X be the total bearing life in hours. The numbers of cycles for the variable forces are:

```
syms X
n1 = X * t1 * nr1 * 60;
n2 = X * t2 * nr2 * 60;
n3 = X * t3 * nr3 * 60;
n4 = X * t4 * nr4 * 60;
% n1 = 15000*X (cycles)
% n2 = 18000*X (cycles)
% n3 = 22500*X (cycles)
% n4 = 7500*X (cycles)
```

The Miner rule is

$$\frac{n_1}{N_1} + \frac{n_2}{N_2} + \frac{n_3}{N_3} + \frac{n_4}{N_4} = 1,$$

where $N_i = L_i$, $i = 1, 2, 3$, or with MATLAB:

```
eq = n1/L1+n2/L2+n3/L3+n4/L4-1;
B =  eval(solve(eq, X));
% bearing life = 1012.497 (h)
```

References

[1] E.A. Avallone, T. Baumeister, A. Sadegh, Marks' Standard Handbook for Mechanical Engineers, 11th edition, McGraw-Hill Education, New York, 2007.
[2] A. Bedford, W. Fowler, Dynamics, Addison Wesley, Menlo Park, CA, 1999.
[3] A. Bedford, W. Fowler, Statics, Addison Wesley, Menlo Park, CA, 1999.
[4] R. Budynas, K.J. Nisbett, Shigley's Mechanical Engineering Design, 9th edition, McGraw-Hill, New York, 2013.
[5] J.A. Collins, H.R. Busby, G.H. Staab, Mechanical Design of Machine Elements and Machines, 2nd edition, John Wiley & Sons, 2009.

[6] A. Ertas, J.C. Jones, The Engineering Design Process, John Wiley & Sons, New York, 1996.

[7] A.S. Hall, A.R. Holowenko, H.G. Laughlin, Schaum's Outline of Machine Design, McGraw-Hill, New York, 2013.

[8] B.G. Hamrock, B. Jacobson, S.R. Schmid, Fundamentals of Machine Elements, McGraw-Hill, New York, 1999.

[9] R.C. Juvinall, K.M. Marshek, Fundamentals of Machine Component Design, 5th edition, John Wiley & Sons, New York, 2010.

[10] K. Lingaiah, Machine Design Databook, 2nd edition, McGraw-Hill Education, New York, 2003.

[11] D.B. Marghitu, Mechanical Engineer's Handbook, Academic Press, San Diego, CA, 2001.

[12] D.B. Marghitu, M.J. Crocker, Analytical Elements of Mechanisms, Cambridge University Press, Cambridge, 2001.

[13] D.B. Marghitu, Kinematic Chains and Machine Component Design, Elsevier, Amsterdam, 2005.

[14] D.B. Marghitu, M. Dupac, N.H. Madsen, Statics with MATLAB, Springer, New York, NY, 2013.

[15] D.B. Marghitu, M. Dupac, Advanced Dynamics: Analytical and Numerical Calculations with MATLAB, Springer, New York, NY, 2012.

[16] D.B. Marghitu, Mechanisms and Robots Analysis with MATLAB, Springer, New York, NY, 2009.

[17] C.R. Mischke, Prediction of stochastic endurance strength, Transaction of ASME, Journal Vibration, Acoustics, Stress, and Reliability in Design 109 (1) (1987) 113–122.

[18] R.L. Mott, Machine Elements in Mechanical Design, Prentice-Hall, Upper Saddle River, NJ, 1999.

[19] W.A. Nash, Strength of Materials, Schaum's Outline Series, McGraw-Hill, New York, 1972.

[20] R.L. Norton, Machine Design, Prentice-Hall, Upper Saddle River, NJ, 1996.

[21] R.L. Norton, Design of Machinery, McGraw-Hill, New York, 1999.

[22] W.C. Orthwein, Machine Component Design, West Publishing Company, St. Paul, 1990.

[23] D. Planchard, M. Planchard, SolidWorks 2013 Tutorial with Video Instruction, SDC Publications, 2013.

[24] C.A. Rubin, The Student Edition of Working Model, Addison–Wesley Publishing Company, Reading, MA, 1995.

[25] A.S. Seireg, S. Dandage, Empirical design procedure for the thermodynamic behavior of journal bearings, Journal of Lubrication Technology 104 (1982) 135–148.

[26] I.H. Shames, Engineering Mechanics – Statics and Dynamics, Prentice-Hall, Upper Saddle River, NJ, 1997.

[27] J.E. Shigley, C.R. Mischke, Mechanical Engineering Design, McGraw-Hill, New York, 1989.

[28] J.E. Shigley, C.R. Mischke, R.G. Budynas, Mechanical Engineering Design, 7th edition, McGraw-Hill, New York, 2004.

[29] J.E. Shigley, J.J. Uicker, Theory of Machines and Mechanisms, McGraw-Hill, New York, 1995.

[30] A.C. Ugural, Mechanical Design, McGraw-Hill, New York, 2004.

[31] J. Wileman, M. Choudhury, I. Green, Computation of member stiffness in bolted connections, Journal of Machine Design 193 (1991) 432–437.

[32] S. Wolfram, Mathematica, Wolfram Media/Cambridge University Press, Cambridge, 1999.

[33] Fundamentals of Engineering. Supplied-Reference Handbook, National Council of Examiners for Engineering and Surveying (NCEES), Clemson, SC, 2001.

[34] MatWeb – material property data, http://www.matweb.com/.

CHAPTER FIVE

Lubrication and sliding bearings

5.1. Introduction

To minimize wear, friction, and temperature among components in motion, a lubricant is introduced. Fig. 5.1A shows a moving part having area A and velocity V. Between the fixed and moving surfaces there is a film of lubricant. The velocity of the film of lubricant in touch with the fixed surface is zero, and the velocity of the lubricant in touch with the moving surface is V. The shear stress in the fluid is

$$\tau = \frac{F}{A} = \mu \frac{\partial v}{\partial y},$$

(5.1)

where F is the force that acts on the lubricant, y is the distance from the fixed surface, $v(y)$ is the velocity, and μ is the *absolute viscosity* or the *dynamic viscosity*. If the gradient of the velocity is constant, as depicted in Fig. 5.1B, the shear stress is

$$\tau = \mu \frac{V}{s}.$$

(5.2)

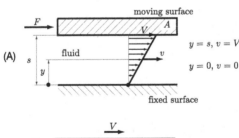

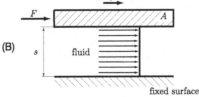

Figure 5.1 Lubrication film in a slider bearing.

Machine component analysis with MATLAB®
https://doi.org/10.1016/B978-0-12-804229-8.00010-4

The absolute viscosity, μ, is specified in SI in newton second per square meter (N s/m^2) or pascal seconds (Pa s). For the US Customary units, the absolute viscosity is measured in pound second per square inch (lb s/in^2) or reyn and

$$1 \text{ reyn} = 1 \text{ lb s/in}^2 = 6890 \text{ N s/m}^2 = 6890 \text{ Pa s.}$$

The absolute viscosity is also commonly measured in microreyn (μreyn) and millipascal second (mPa s). Another unit for the absolute viscosity is the centipoise, cp, given by

$$1 \text{ cp} = 1 \text{ mPa s.}$$

A lubricant with greater μ corresponds to a thicker lubricant.

The *kinematic viscosity*, v, is by definition

$$v = \frac{\mu}{\rho}, \qquad (5.3)$$

where μ is the absolute viscosity and ρ is the mass density. The kinematic viscosity is measured in cm^2/s or stoke (St):

$$1 \text{ m}^2/\text{s} = 10^4 \text{ St,}$$
$$1 \text{ cSt (centistoke)} = 1 \text{ mm}^2/\text{s.}$$

The mass density for various temperatures is [9]

$$\rho = 0.89 - 0.00063\,(°\text{C} - 15.6) \text{ g/cm}^3, \qquad (5.4)$$

or

$$\rho = 0.89 - 0.00035\,(°\text{F} - 60) \text{ g/cm}^3. \qquad (5.5)$$

The kinematic viscosity of oils, v, is measured with the Saybolt Universal Viscometer, as shown in Fig. 5.2. With the Saybolt Universal Viscometer one can determine the time, in Saybolt seconds, necessary for a given lubricant to flow through an orifice. Using the Saybolt viscometer, the absolute viscosity can be determined [9] as

$$\mu = \left(0.22\,t - \frac{180}{t}\right)\rho \qquad \text{mPa s or cp,} \qquad (5.6)$$

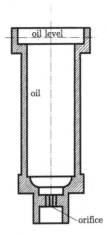

Figure 5.2 Saybolt universal viscometer.

and

$$\mu = 0.145 \left(0.22\, t - \frac{180}{t} \right) \rho \qquad\qquad \text{µreyn,} \qquad\qquad (5.7)$$

where t is the time in seconds and ρ is in g/cm^3.

The classification of the oils based on viscosity is done by the Society of Automotive Engineers (SAE). The SAE reports only the viscosity of the oil, and it is not correlated with the performance. There are two different SAE viscosities: for monograde oils (single viscosity oils) and multigrade oils with the letter W (Winter). The kinematic viscosity of monograde oils, SAE 10 or SAE 15, is given at 100°C (212°F) in cSt. The numbers 10 or 15 define the kinematic viscosity at 100°C. The SAE 10 is a less viscous oil than an SAE 30 oil. The multigrade oils are multi-viscosity oils and have two parts, e.g., SAE 15W-40 or SAE 20W-50. For a multigrade oil, the viscosity at low temperatures is given by the first number (15W or 20W) and the kinematic viscosity at 100°C is provided by the second number. The low temperature dynamic viscosity is expressed in cp. For W temperature, the viscosity of oil is lower for a lower W number. For the multigrade oils SAE 10W-30 and SAE 15W-30, the number 30 specifies that oils have the same kinematic viscosity at 100°C. For low temperature (W) the 10W-30 oil is a less viscous oil than the SAE 15W-30 (10 < 15). A synthetic oil should be considered for better performance at extreme temperatures and special applications.

An equation for the approximate viscosity versus temperature, using curve fits, is given by Seireg and Dandage [27] as

$$\mu = \mu_0\, e^{b/(T+95)},\qquad(5.8)$$

where μ is in µreyn, T is in °F, and the values of μ_0 and b are given by

$$\mu_0 = 0.0158\ \text{µreyn and}\ b = 1157.5\,°F\ \text{for SAE 10},$$
$$\mu_0 = 0.0136\ \text{µreyn and}\ b = 1271.6\,°F\ \text{for SAE 20},$$
$$\mu_0 = 0.0141\ \text{µreyn and}\ b = 1360.0\,°F\ \text{for SAE 30},$$
$$\mu_0 = 0.0121\ \text{µreyn and}\ b = 1474.4\,°F\ \text{for SAE 40},$$
$$\mu_0 = 0.0170\ \text{µreyn and}\ b = 1509.6\,°F\ \text{for SAE 50},$$
$$\mu_0 = 0.0187\ \text{µreyn and}\ b = 1564.0\,°F\ \text{for SAE 60}.$$

5.2. Petroff's law

Hydrodynamic lubrication does not depend on the introduction of the lubricant under pressure. The pressure of the lubricant is created by the surface motion. Fig. 5.3 depicts a bearing with hydrodynamic lubrication and a small radial load. The axial lubricant flow is neglected. The eccentricity between the journal bearing and the shaft is negligible. Fig. 5.3 shows the shaft with radius R, length of the bearing L, and radial clearance c. The surface velocity is $V = 2\pi R n$, where n (rev/s) is the angular speed of the

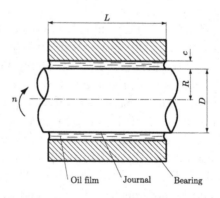

Figure 5.3 Hydrodynamic bearing.

shaft. The shear stress is given by Eq. (5.2) which could also be written as

$$\tau = \mu \frac{V}{s} = \frac{2\pi R \mu n}{c}.$$ (5.9)

The force necessary to shear the film is

$$F = \tau A,$$

where $A = 2\pi RL$. The friction torque is calculated with the formula

$$T_f = FR = (\tau A)\, R = \left(\frac{2\pi R \mu n}{c} 2\pi RL \right) R = \frac{4\pi^2 \mu n L R^3}{c}.$$ (5.10)

The radial load per unit of projected bearing area is

$$P = \frac{W}{2RL},$$

where W is a small radial load on the bearing The friction torque is

$$T_f = fWR = f\,(2RLP)\, R = 2R^2 fLP,$$ (5.11)

where f is the coefficient of friction. Using Eqs. (5.10) and (5.11), the coefficient of friction is

$$f = 2\pi^2 \left(\frac{\mu n}{P} \right) \left(\frac{R}{c} \right).$$ (5.12)

Equation (5.12) is the Petroff's equation, or Petroff's law. The dimensionless variables, $\left(\frac{\mu n}{P} \right)$ and $\left(\frac{R}{c} \right)$, represent bearing parameters. The symbolical expression for Petroff's equation obtained with MATLAB® is:

```
syms R L c n mu P f
V = 2*pi*R*n;
tau = mu*V/c;
A = 2*pi*R*L;
F = tau*A;
Tf = F*R;
fprintf('Tf = %s \n',Tf)
% P = W/(2*R*L);
W = P*(2*R*L);
% Tf = f*W*R
```

```
f = Tf/(W*R);
fprintf('f = %s \n',f)
% Tf = (4*L*R^3*mu*n*pi^2)/c
% f = (2*R*mu*n*pi^2)/(P*c)
```

The Sommerfeld number S, or the *characteristic number* for hydrodynamic bearings, is given by

$$S = \frac{\mu n}{P} \left(\frac{R}{c}\right)^2,$$ (5.13)

where μ is the absolute viscosity (reyn), n is the angular speed (rev/s), P is the pressure or radial load per unit of projected bearing area (psi), R is the journal radius (in), and c is the radial clearance (in). The power loss in W and kW is

$$H = 2\pi\, T_f\, n, \quad \text{W},$$

$$H = \frac{T_f\, n}{9549}, \quad \text{kW},$$ (5.14)

where n is the angular shaft speed in rev/s and T_f is the friction torque in N m.

The power loss in hp is

$$H = \frac{T_f n}{5252}, \quad \text{hp},$$ (5.15)

where T_f is the friction torque in lb ft and n the angular speed of the shaft speed in rpm.

5.3. Reynolds equation

Fig. 5.4 shows a small volume element of lubricant film. The lubricant exhibits a Newtonian, incompressible laminar flow and has a constant viscosity. The pressure acts on the $dy\,dz$ surface, and the shear acts on $dy\,dz$. From the force diagram this results in

$$p\, dy\, dz - \tau\, dx\, dz - \left(p + \frac{dp}{dx}dx\right)\, dy\, dz + \left(\tau + \frac{\partial \tau}{\partial y}dy\right)\, dx\, dz = 0,$$ (5.16)

or the pressure gradient being

$$\frac{dp}{dx} = \frac{\partial \tau}{\partial y}.$$ (5.17)

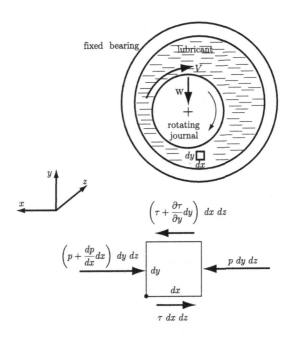

Figure 5.4 Element of lubricant film.

The pressure of the film is a function of the x-coordinate, $p = p(x)$. The shear stress depends on the derivative of the velocity $v(x, y)$ of any particle of lubricant with respect to y and the absolute viscosity, namely $\tau = \mu \dfrac{\partial v(x, y)}{\partial y}$. This results in

$$\frac{dp}{dx} = \mu \frac{\partial^2 v}{\partial^2 y} \quad \text{or} \quad \frac{\partial^2 v}{\partial^2 y} = \frac{1}{\mu} \frac{dp}{dx}. \tag{5.18}$$

Integrating twice with respect to y and keeping x constant, the following relations are obtained:

$$\frac{\partial v}{\partial y} = \frac{1}{\mu}\left(\frac{dp}{dx}y + C_1\right) \tag{5.19}$$

and

$$v = \frac{1}{\mu}\left(\frac{dp}{dx}\frac{y^2}{2} + C_1 x + C_2\right). \tag{5.20}$$

The two integration constants are determined from the boundary conditions: for $y = 0 \Longrightarrow v = 0$, and for $y = s \Longrightarrow v = V$. The velocity field is

$$v = \frac{1}{2\mu} \frac{dp}{dx} (y^2 - sy) + \frac{V}{s} y. \tag{5.21}$$

The MATLAB solution for the differential equations is obtained from:

```
syms dp mu v(y) V s dpx
% dpx = dp/dx
sol = dsolve(diff(v,2)==dpx/mu, v(0)==0, v(s)==V);
fprintf('v = %s \n',simplify(sol))
% v = (y*(2*V*mu - dpx*s^2 + dpx*s*y))/(2*mu*s)
```

The velocity field of the lubricant is obtained from Eq. (5.21), and it is sketched in Fig. 5.5. The volume of lubricant, Q, is calculated with

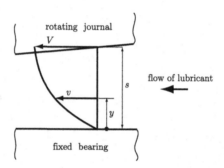

Figure 5.5 Velocity distribution.

$$Q = \int_0^s v(x, y) dy = \frac{Vs}{2} - \frac{s^3}{12\mu} \frac{dp}{dx}, \tag{5.22}$$

or with MATLAB:

```
Q = int(sol,y,0,s);
fprintf('Q = %s \n',Q)
% Q = (V*s)/2 - (dpx*s^3)/(12*mu)
```

For incompressible lubricant $\dfrac{dQ}{dx} = 0$, and, differentiating Eq. (5.22), the Reynolds equation for one-dimensional flow is obtained as

$$\frac{dQ}{dx} = \frac{V}{2}\frac{ds}{dx} - \frac{d}{dx}\left(\frac{s^3}{12\mu}\frac{dp}{dx}\right),$$

or

$$\frac{d}{dx}\left(\frac{s^3}{\mu}\frac{dp}{dx}\right) = 6V\frac{ds}{dx}. \qquad (5.23)$$

For a two-dimensional flow, the Reynolds equation is

$$\frac{\partial}{\partial x}\left(\frac{s^3}{\mu}\frac{\partial p}{\partial x}\right) + \frac{\partial}{\partial z}\left(\frac{s^3}{\mu}\frac{\partial p}{\partial z}\right) = 6V\frac{\partial s}{\partial x}. \qquad (5.24)$$

For analysis and design of short bearings, the following equation is recommended [9]:

$$\frac{\partial}{\partial z}\left(\frac{s^3}{\mu}\frac{\partial p}{\partial z}\right) = 6V\frac{\partial h}{\partial x}. \qquad (5.25)$$

5.4. Design of hydrodynamic bearings

The solutions of Reynolds equation have been changed into charts by Raimondi and Boyd [26,4,9]. Figs. 5.7–5.12 depict the solutions for different bearings. The variables used with the charts are explained in Figs. 5.13, 5.14. The Raimondi and Boyd plots are calculated in terms of the bearing characteristic number, or Sommerfeld variable, S.

From Fig. 5.6, the left limit of the zone gives the minimum film thickness variable for minimum friction and the right limit is the minimum film thickness variable for maximum load. The most favorable values for the Sommerfeld variable are [9]

$$S = \begin{cases} 0.037 & L/D = 1/2 & \text{for min. friction,} \\ 0.35 & L/D = 1/2 & \text{for max. load,} \\ 0.082 & L/D = 1/1 & \text{for min. friction,} \\ 0.21 & L/D = 1/1 & \text{for max. load,} \end{cases} \qquad (5.26)$$

where D is the diameter and L is the length of the bearing.

Using Trumpler's empirical equation, a minimal acceptable oil film thickness, h_o, is calculated with [9]

$$h_o \geq h_{o\,min} = 0.0002 + 0.00004\,D \text{ where } h_o \text{ and } D \text{ in inches,}$$
$$h_o \geq h_{o\,min} = 0.005 + 0.00004\,D \text{ where } h_o \text{ and } D \text{ in mm.} \qquad (5.27)$$

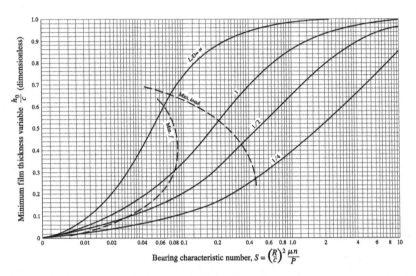

Figure 5.6 Minimum film thickness variable, h_0, [26]. From Budynas–Nisbett: Shigley's Mechanical Engineering Design, Eighth Edition, McGraw-Hill, 2006. Used with permission from McGraw Hill Inc.

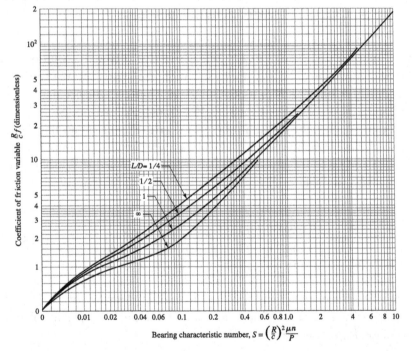

Figure 5.7 Coefficient of friction variable, f [26]. From Budynas–Nisbett: Shigley's Mechanical Engineering Design, Eighth Edition, McGraw-Hill, 2006. Used with permission from McGraw Hill Inc.

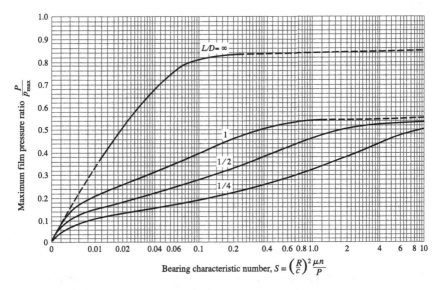

Figure 5.8 Maximum film pressure, $p_{\max}$ [26]. From Budynas–Nisbett: Shigley's Mechanical Engineering Design, Eighth Edition, McGraw-Hill, 2006. Used with permission from McGraw Hill Inc.

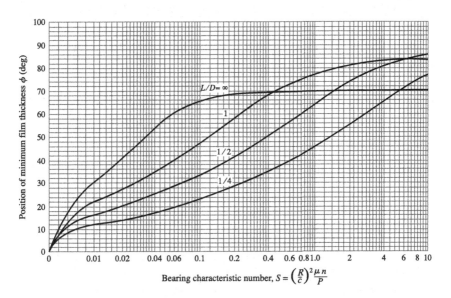

Figure 5.9 Position angle of minimum film thickness, ϕ [26]. From Budynas–Nisbett: Shigley's Mechanical Engineering Design, Eighth Edition, McGraw-Hill, 2006. Used with permission from McGraw Hill Inc.

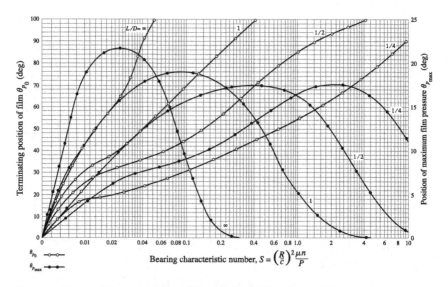

Figure 5.10 Terminating position of the lubricant film, θ_{p0}, and the position of maximum film pressure, $\theta_{p_{max}}$ [26]. From Budynas–Nisbett: Shigley's Mechanical Engineering Design, Eighth Edition, McGraw-Hill, 2006. Used with permission from McGraw Hill Inc.

Trumpler suggests a safety factor of 2 for the load W, and doubling P reduces S by half.

Juvinall and Marshek present some representative unit sleeve bearing load,

$$P = \frac{W_{max}}{LD},$$

(5.28)

for design use [9]:

(a) for relatively steady loads

$$P = \begin{cases} 0.8\text{–}1.5 \text{ MPa or } 120\text{–}250 \text{ psi} & \text{for electric motors,} \\ 1.0\text{–}2.0 \text{ MPa or } 150\text{–}300 \text{ psi} & \text{for steam turbines,} \\ 0.8\text{–}1.5 \text{ MPa or } 120\text{–}250 \text{ psi} & \text{for gear reducers,} \\ 0.6\text{–}1.2 \text{ MPa or } 100\text{–}180 \text{ psi} & \text{for centrifugal pumps.} \end{cases}$$

(5.29)

(b) for diesel engines and rapidly fluctuating loads

$$P = \begin{cases} 6\text{–}12 \text{ MPa or } 900\text{–}1700 \text{ psi} & \text{for main bearings,} \\ 8\text{–}15 \text{ MPa or } 1150\text{–}2300 \text{ psi} & \text{for connecting rod bearings.} \end{cases}$$

(5.30)

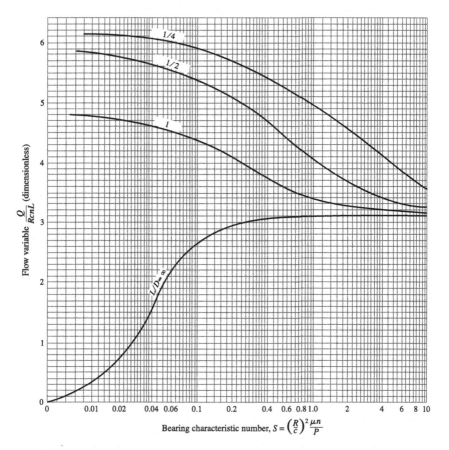

Figure 5.11 Volumetric oil-flow rate into the bearing, Q [26]. From Budynas–Nisbett: Shigley's Mechanical Engineering Design, Eighth Edition, McGraw-Hill, 2006. Used with permission from McGraw Hill Inc.

(c) for automotive gasoline engines and rapidly fluctuating loads

$$P = \begin{cases} 4\text{--}5 \text{ MPa or } 600\text{--}750 \text{ psi} & \text{for main bearings,} \\ 10\text{--}15 \text{ MPa or } 1700\text{--}2300 \text{ psi} & \text{for connecting rod bearings.} \end{cases} \quad (5.31)$$

For the clearance ratio, c/R, the maximum temperature, T, and the bearing ratio, L/D, the following values are recommended for design [9]:

$$c/R = \begin{cases} 0.001 & \text{for } 25 < D < 150 \text{ mm and precision bearings,} \\ < 0.002 & \text{for general machinery bearings,} \\ < 0.004 & \text{for rough-service machinery,} \end{cases} \quad (5.32)$$

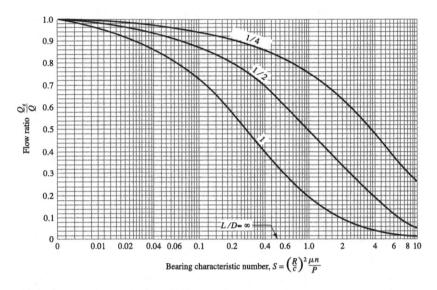

Figure 5.12 Volumetric side-flow leakage, Q_S [26]. From Budynas–Nisbett: Shigley's Mechanical Engineering Design, Eighth Edition, McGraw-Hill, 2006. Used with permission from McGraw Hill Inc.

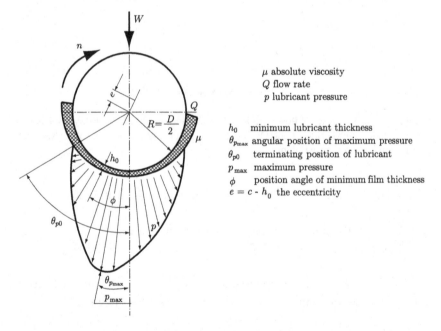

Figure 5.13 Notation for Raimondi and Boyd charts [26]. From Budynas–Nisbett: Shigley's Mechanical Engineering Design, Eighth Edition, McGraw-Hill, 2006. Used with permission from McGraw Hill Inc.

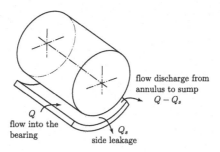

Figure 5.14 Lubricant flow.

$$T = \begin{cases} < (90\,^\circ\text{C--}122\,^\circ\text{C}), \\ < (200\,^\circ\text{F--}250\,^\circ\text{F}), \end{cases} \tag{5.33}$$

$$L/D = \begin{cases} 0.25\text{--}0.75 & \text{frequently,} \\ 1 & \text{for older machinery.} \end{cases} \tag{5.34}$$

5.5. Examples

Example 5.1. An oil-lubricated journal bearing has diameter $D = 2$ in, length $L = 1$ in, and diametral clearance $c_d = 0.0012$ in. The angular speed of the shaft is $n = 300$ rpm. The bearing is lubricated by SAE 30 oil at an average temperature of $T = 175\,^\circ\text{F}$. Using Petroff's equation, estimate the power loss and friction torque.

Solution

The assumptions for using the Petroff's equation are: the eccentricity between the journal bearing and journal is negligible; there is no lubricant flow in the axial direction; the radial load is small. The input data for the problem are:

```
D = 2;        % (in)
L = 1;        % (in)
cd = 0.0012;  % (in)
n = 300;      % (rpm)
T = 175;      % (degF)
R = D/2;      % radius of the bearing
c = cd/2;     % radial clearance
ns = n/60;    % (rps)
```

The viscosity of oil for SAE 30 is calculated with Eq. (5.8):

```
% SAE 30
mu0 = 0.0141; % (micro reyn)
b = 1360.0;   % (degF)
mu = mu0*exp(b/(T+95)); % (micro reyn)
% mu =  2.172 (micro reyn)
mu = mu*10^(-6); % (reyn)
```

The friction torque is obtained from Eq. (5.10):

```
Tf = 4 * pi^2 * mu * ns * L * R^3 /c;
% Tf =  0.714 (lb in)
Tf = Tf/12;
% Tf =  0.060 (lb ft)
```

The power loss is calculated with Eq. (5.15):

```
H = n * Tf /5252;
% H =  0.003 (hp)
```

Example 5.2. A lubricated journal bearing has diameter $D = 2.5$ in, length $L = 2.5$ in, and clearance $c = 0.002$ in. The shaft rotates with the angular speed of $n = 1550$ rpm. The bearing is lubricated by an SAE 20 oil at an average controlled temperature of $T = 120\,°F$. For a radial load of $W = 950$ lb, estimate the coefficient of friction and the power loss.

Solution

The Petroff's approach is suitable: the eccentricity between the shaft and bearing is negligible, and the flow in the axial direction is negligible. The input data for the problem are:

```
D = 2.5;     % (in)
L = 2.5;     % (in)
c = 0.002;   % (in)
n = 1550;    % (rpm)
T = 120;     % (degF)
W = 950;     % (lb)
R = D/2;     % radius of the bearing
ns = n/60;   % (rps)
```

The viscosity, μ, of the SAE 20 oil at the given temperature is:

```
muO = 0.0136; % (micro reyn)
b = 1271.6;   % (degF)
mu = muO*exp(b/(T+95)); % (micro reyn)
mu = mu*10^(-6); % (reyn)
% mu = 5.037e-06 (reyn)
```

The load intensity is:

```
% load intensity
P = W/(L*D);
% P = 152.000 (lb/in^2)
```

The coefficient of friction is obtained with:

```
f = 2 * pi^2 * (mu * ns/P) * (R/c);
% f = 0.0106
```

The friction torque is calculated with:

```
Tf = f * W * R;
% Tf = 12.541 (lb in)
Tf = Tf/12;
% Tf =  1.045 (lb ft)
```

The power loss is

```
H = n*Tf/5252;
% H = 0.3084 (hp)
```

Next the coefficient of friction is checked using the Raimondi and Boyd chart shown in Fig. 5.7. The bearing characteristic number is:

```
Sc = (mu * ns/P) * (R/c)^2;
% Sc =  0.334
```

From Fig. 5.7 two vectors are created for the characteristic number, S, and the corresponding Rf/c, for $L/D = 1$ [9,10]:

```
% L/D = 1/1
S   = [0.0062 0.0248 0.08 0.0993 0.21 0.397 0.632 1.59]; % S
Rfc = [0.53   1.22   2.4  2.7    4.8  8.2   13    32 ]; % R f/c
```

The previous discrete data is interpolated:

```
ffc  = csapi( S, Rfc );
```

The value of the interpolated function at $Sc = 0.334$ is:

```
fc = fnval(ffc , Sc);
% R f/c =   7.073
```

and the coefficient of friction is:

```
fs = fc*c/R;
% fs = 0.0113
```

The relative error for the coefficient of friction is:

```
err = (fs-f)/fs;
% relative error = 0.0669
```

Example 5.3. A bearing with hydrodynamic lubrication has a journal diameter $D = 0.2$ m, length $L = 0.1$ m, and radial clearance $c = 0.15$ mm, as shown in Fig. 5.15. The angular speed of the shaft is $n = 900$ rpm. A constant load of $W = 5000$ N is applied to the bearing. The average temperature of the SAE 20 oil is $t_{avg} = 70 °C$. Find the minimum oil film thickness, h_0, bearing coefficient of friction, f, total oil flow rate, Q, and flow ratio (side flow/total flow) Q_s/Q.

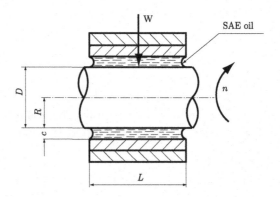

Figure 5.15 Example 5.3.

Solution
The given input for the problem is:

```
D = 0.200;       % (m)
L = 0.100;       % (m)
c = 0.15*10^-3;  % (m)
n = 900;         % (rpm)
```

```
t = 70;          % (degC)
W = 5*10^3;      % (N)
R = D/2;         % radius of the bearing
ns = n/60;       % (rps)
```

The dynamic viscosity is calculated with:

```
% T(F) = T(C)*9/5 + 32
T = 9*t/5+32;    % (degF)
% SAE 20
mu0 = 0.0136;  % (micro reyn)
b = 1271.6;      % (degF)
mu = mu0*exp(b/(T+95)); % (micro reyn)
mu = mu*10^(-6); % (reyn)
% 1 reyn = 6890 Pa s
mu = mu*6890;
% mu = 2.072e-06 (reyn)
% mu = 0.014     (Pa s) or (N s/m^2)
```

The pressure is:

```
P = W/(L*D);
% P = 2.500e+05 (N/m^2)
```

and the Sommerfeld number is:

```
Sc = (mu * ns/P) * (R/c)^2;
% Sc =  0.381
```

For $L/D = 1/2$, the following vectors are introduced:

• the Sommerfeld number vector

```
S = ...
[0.021 0.030 0.036 0.041 0.057 0.085 0.140 0.274 0.342 0.428 0.762 1.714];
```

• the corresponding vector for the minimum film thickness variable, h_0/c

```
hoc = ...
[0.08 0.10 0.11 0.12 0.15 0.195 0.26 0.37 0.425 0.47 0.59 0.76];
```

• the corresponding vector for the coefficient of friction variable, Rf/c

```
Rfc = ...
[1.3 1.6 1.75 1.9 2.3 3.1 4.4 7.3 8.7 10 16 36];
```

• the corresponding vector for the volumetric oil-flow rate, $Q/(RcnL)$

```
Qc = ...
[5.75 5.7 5.65 5.6 5.55 5.45 5.25 4.95 4.8 4.65 4.3 3.8];
```

• the corresponding vector for side leakage flow/total flow, Q_s/Q

```
Qsc = ...
[0.95 0.94 0.93 0.92 0.91 0.88 0.84 0.76 0.72 0.68 0.56 0.37];
```

To interpolate the data, the MATLAB function `csapi(X,Y)` is used:

```
fhoc = csapi( S, hoc );
ffc  = csapi( S, Rfc );
fQR  = csapi( S, Qc  );
fQs  = csapi( S, Qsc );
```

The values of the interpolated data functions at *Sc* are calculated with:

```
% Sc = 0.381
hc = fnval(fhoc, Sc);
fc = fnval(ffc, Sc);
Qc = fnval(fQR, Sc);
Qsc= fnval(fQs, Sc);
% ho/c =   0.448
% R f/c =   9.327
% Q/(R c n L) =   4.726
% Qs/Q =   0.701
```

The minimum film thickness, h_o, coefficient of friction f, total flow Q, and side leakage flow Q_s are found with:

```
h = c*hc;
f = c*fc/R;
Q = Qc*R*c*n*L;
Qs = Q*Qsc;
% h = 6.724e-05 (m)
% f =   0.014
% Q = 6.381e-03 (m^3/s)
% Qs = 4.472e-03 (m^3/s)
```

From the total flow $Q = 6.381\,(10^{-3})$ m^3/s, the side flow $Q_s = 4.472\,(10^{-3})$ m^3/s pours out the ends. The rest of the flow is recirculated.

Example 5.4. An oil-lubricated bearing of an electric motor has a journal that rotates with $n = 250$ rpm. The radial load on the bearing is $W = 1000$ lb. Determine an appropriate bearing length, radius, clearance, lubricant, and average oil temperature.

Solution

The MATLAB initial data are:

```
W = 1000;    % (lb)
n =  250;    % (rpm)
ns = n/60;   % (rps)
```

To reduce the radial space, the bearing ratio $L/D = 1$ is selected. From Eq. (5.29), for electric motors the range is $P \in [120, 250]$ psi. Due to the equation for the unit load $P = W/(LD)$, the length of the bearing should be $L \in [L_1, L_2]$ where:

```
P1 = 120; % (psi)
P2 = 250; % (psi)
L1 = sqrt(W/P1);
L2 = sqrt(W/P2);
% L1 =  2.887 (in)
% L2 =  2.000 (in)
```

For an arbitrarily selected $L = 2.5$ in, the intensity P is:

```
L = 2.5; % (in)
D = L;
P = W/(L*D);
% P = 160.000 (lb/in^2),(psi)
```

The clearance ratio is selected as $c/R = 0.0015$ for general machinery bearings, and then the clearance is:

```
cR = 0.0015;
R = L/2;
c = cR * R;
% c = 0.0019 (in)
```

For $L/D = 1$, the optimum operating ranges from Fig. 5.6 are at $S = 0.082$, for minimum friction, and $S = 0.21$, for maximum load. To find the viscosity, we assume a mid-optimum range:

```
Smin = 0.082;      % S for min friction
Smax = 0.21 ;      % S for max load
S = (Smin+Smax)/2; % mid-optimum range
% S =  0.146
mu = S / ((ns/P) * (R/c)^2);
% mu = 1.261e-05 (reyn)
```

The oil used for lubrication is SAE 20, and the average temperature is calculated with:

```
mu = mu*10^6; % micro reyn
% SAE 20
mu0 = 0.0136; % micro reyn
b = 1271.6;   % degF
T = b/log(mu/mu0) - 95;
% T = 91.110 (degF)
```

Next the clearance is checked with Trumpler empirical equation $h_o \geq$ $h_{o\,min} = 0.0002 + 0.00004D$:

```
homin = 0.0002+0.00004*D;
% homin = 3.000e-04 (in)
```

The minimum film thickness, h_o, is calculated using a safety factor of $C_s=2$ applied to the load, $W = C_s\,W$. Doubling the intensity P, the bearing characteristic number, S, is reduced by half. The minimum oil film thickness, h_o, is calculated from the ratio, h_o/c:

```
Cs = 2;
W = Cs * W;
% doubling P reduces S by half
Sn = S/Cs;
% Sn =  0.073
% L/D = 1/1
S = ...
[0.0062 0.0248 0.08 0.041 0.0993 0.21 0.397 0.614];
% ho/c
hoc = ...
[0.03   0.13   0.30 0.19 0.345  0.54 0.69 0.78];
% interpolate the data
fhoc = csapi( S, hoc );
hc = fnval(fhoc, Sn); % ho/c
```

```
% ho/c = 0.282
ho = c * hc;
% ho = 5.297e-04 (in)
    if ho > homin
fprintf('the criterion is satisfied \n');
    else
fprintf('the criterion is not satisfied \n');
    end
```

The Trumpler criterion is satisfied.

References

[1] E.A. Avallone, T. Baumeister, A. Sadegh, Marks' Standard Handbook for Mechanical Engineers, 11th edition, McGraw-Hill Education, New York, 2007.

[2] A. Bedford, W. Fowler, Dynamics, Addison Wesley, Menlo Park, CA, 1999.

[3] A. Bedford, W. Fowler, Statics, Addison Wesley, Menlo Park, CA, 1999.

[4] R. Budynas, K.J. Nisbett, Shigley's Mechanical Engineering Design, 9th edition, McGraw-Hill, New York, 2013.

[5] J.A. Collins, H.R. Busby, G.H. Staab, Mechanical Design of Machine Elements and Machines, 2nd edition, John Wiley & Sons, 2009.

[6] A. Ertas, J.C. Jones, The Engineering Design Process, John Wiley & Sons, New York, 1996.

[7] A.S. Hall, A.R. Holowenko, H.G. Laughlin, Schaum's Outline of Machine Design, McGraw-Hill, New York, 2013.

[8] B.G. Hamrock, B. Jacobson, S.R. Schmid, Fundamentals of Machine Elements, McGraw-Hill, New York, 1999.

[9] R.C. Juvinall, K.M. Marshek, Fundamentals of Machine Component Design, 5th edition, John Wiley & Sons, New York, 2012.

[10] R.C. Juvinall, K.M. Marshek, Fundamentals of Machine Component Design – Instructor's Manual, John Wiley & Sons, New York, 2012.

[11] K. Lingaiah, Machine Design Databook, 2nd edition, McGraw-Hill Education, New York, 2003.

[12] D.B. Marghitu, Mechanical Engineer's Handbook, Academic Press, San Diego, CA, 2001.

[13] D.B. Marghitu, M.J. Crocker, Analytical Elements of Mechanisms, Cambridge University Press, Cambridge, 2001.

[14] D.B. Marghitu, Kinematic Chains and Machine Component Design, Elsevier, Amsterdam, 2005.

[15] D.B. Marghitu, M. Dupac, N.H. Madsen, Statics with MATLAB, Springer, New York, NY, 2013.

[16] D.B. Marghitu, M. Dupac, Advanced Dynamics: Analytical and Numerical Calculations with MATLAB, Springer, New York, NY, 2012.

[17] D.B. Marghitu, Mechanisms and Robots Analysis with MATLAB, Springer, New York, NY, 2009.

[18] C.R. Mischke, Prediction of stochastic endurance strength, Transaction of ASME, Journal Vibration, Acoustics, Stress, and Reliability in Design 109 (1) (1987) 113–122.

[19] R.L. Mott, Machine Elements in Mechanical Design, Prentice-Hall, Upper Saddle River, NJ, 1999.

[20] W.A. Nash, Strength of Materials, Schaum's Outline Series, McGraw-Hill, New York, 1972.

[21] R.L. Norton, Machine Design, Prentice-Hall, Upper Saddle River, NJ, 1996.

[22] R.L. Norton, Design of Machinery, McGraw-Hill, New York, 1999.

[23] W.C. Orthwein, Machine Component Design, West Publishing Company, St. Paul, 1990.

[24] D. Planchard, M. Planchard, SolidWorks 2013 Tutorial with Video Instruction, SDC Publications, 2013.

[25] C.A. Rubin, The Student Edition of Working Model, Addison-Wesley Publishing Company, Reading, MA, 1995.

[26] A.A. Raimondi, J. Boyd, A solution for the finite journal bearing and its application to analysis and design, Part I, II, and III, ASLE Transactions 1 (1) (1958) 159–209, in Lubrication Science and Technology, Pergamon Press, New York.

[27] A.S. Seireg, S. Dandage, Empirical design procedure for the thermodynamic behavior of journal bearings, Journal of Lubrication Technology 104 (1982) 135–148.

[28] I.H. Shames, Engineering Mechanics – Statics and Dynamics, Prentice-Hall, Upper Saddle River, NJ, 1997.

[29] J.E. Shigley, C.R. Mischke, Mechanical Engineering Design, McGraw-Hill, New York, 1989.

[30] J.E. Shigley, C.R. Mischke, R.G. Budynas, Mechanical Engineering Design, 7th edition, McGraw-Hill, New York, 2004.

[31] J.E. Shigley, J.J. Uicker, Theory of Machines and Mechanisms, McGraw-Hill, New York, 1995.

[32] A.C. Ugural, Mechanical Design, McGraw-Hill, New York, 2004.

[33] J. Wileman, M. Choudhury, I. Green, Computation of member stiffness in bolted connections, Journal of Machine Design 193 (1991) 432–437.

[34] S. Wolfram, Mathematica, Wolfram Media/Cambridge University Press, Cambridge, 1999.

[35] Fundamentals of Engineering Supplied-Reference Handbook, National Council of Examiners for Engineering and Surveying (NCEES), Clemson, SC, 2001.

[36] MatWeb – material property data, http://www.matweb.com/.

CHAPTER SIX

Spur gears

6.1. Introduction

The standards for gear are specified according to American Gear Manufacturers Association (AGMA). The spur gears are used to shift the motion between parallel shafts. The gears are durable and have very good efficiency (up to 98%). Two gears in contact are shown in Fig. 6.1. For the gears in contact, the common normal to the surfaces at the point of contact intersects the line of gear centers at the same point P, called the pitch point, as seen in Fig. 6.1. This is defined as the law of conjugate gear-tooth action and satisfies the constant condition for the angular speed ratios. The involute curve complies with the conjugate gear-tooth action basic law. The involute of a circle (base circle) of radius r_b is given by the Cartesian parametric equations

$$x = (\cos\theta + \theta\sin\theta)\, r_b,$$
$$y = (\sin\theta - \theta\cos\theta)\, r_b,$$

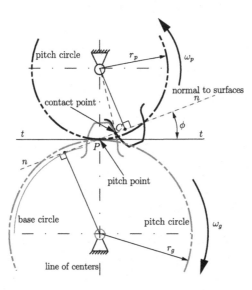

Figure 6.1 Two gears in contact. From Dan B. Marghitu, Kinematic Chains and Machine Elements Design, Elsevier, 2005.

Machine component analysis with MATLAB®
https://doi.org/10.1016/B978-0-12-804229-8.00011-6

where θ is the angle in radians, or in MATLAB®

```
syms rb theta
rb = 0.1;
theta = 0:pi/100:2*pi;
%  circle of radius rb
xc = rb*cos(theta);
yc = rb*sin(theta);
% the parametric equation of the involute
xi = (cos(theta) + theta.*sin(theta))*rb;
yi = (sin(theta) - theta.*cos(theta))*rb;
figure
plot(xc,yc,'-r', xi,yi,'-b','linewidth',2)
grid on
axis equal
```

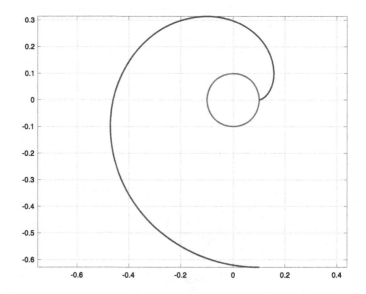

Figure 6.2 Involute of a base circle of radius $r_b = 0.1$.

Fig. 6.2 shows the involute of a base circle with radius $r_b = 0.1$. The pitch circles are the imaginary tangent circles of the gears in contact. The contact point for the pitch circles is the pitch point P. For two connected gears, the smaller of the two is the pinion and the larger is the gear, while the angular speed ratio is defined as

$$i = \frac{\omega_p}{\omega_g} = -\frac{d_g}{d_p} \quad \text{or} \quad i_{12} = \frac{\omega_1}{\omega_2} = -\frac{d_2}{d_1}, \tag{6.1}$$

where d is the pitch diameter, ω represents the angular speed expressed in rad/s, and, when used, symbol n represents the angular speed in rpm. The minus sign in Eq. (6.1) shows that the two gears are rotating in opposite directions. The center distance between the gears defined as

$$c = \frac{d_g + d_p}{2} = r_g + r_p, \tag{6.2}$$

where the radius of the pitch circle is calculated as $r = d/2$. For the gears shown in Fig. 6.2, line tt is the common tangent to the pitch circles, and line nn is the common normal line to the surfaces at the contact point. The pressure angle, ϕ, is the angle of the line nn with the line tt and has the standard value of 20° for English and SI units. The pressure angle $\phi = 25°$ is a standard value in the US as well. The outer circle of a gear is the addendum circle and its radius is

$$r_a = a + r, \quad \text{or} \quad r_{ap} = a + r_p \quad \text{and} \quad r_{ag} = a + r_g,$$

where a is the addendum shown in Fig. 6.3.

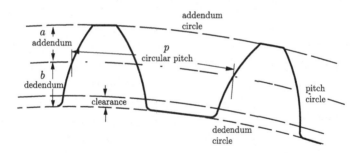

Figure 6.3 Gear teeth.

The external teeth profiles are reduced from the pitch circle, a length b named dedendum. The circular pitch, p, is measured in mm (SI units) or in inches (English units) and is given by

$$p = \frac{\pi d}{N} \quad \text{or} \quad p = \frac{\pi d_p}{N_p} = \frac{\pi d_g}{N_g}, \tag{6.3}$$

where N_p and N_g represent the number of teeth of the pinion and gear, respectively.

For the English units the diametral pitch, P_d, is

$$P_d = \frac{N}{d} \quad \text{or} \quad P_d = \frac{N_p}{d_p} = \frac{N_g}{d_g}. \tag{6.4}$$

If p is in inches and P_d is in teeth per inch then $p\,P_d = \pi$.

The module, m, is used only with SI in mm and is given by

$$m = \frac{d}{N} \quad \text{or} \quad m = \frac{d_p}{N_p} = \frac{d_g}{N_g}. \tag{6.5}$$

If p is in mm and m is in mm then $p/m = \pi$.

For the full-depth involute teeth with $\phi = 20°$, the addendum standard value and the dedendum minimum value in SI units are $a = m$ and $b = 1.25\,m$, respectively.

For English units the addendum and the minimum dedendum are [7] as follows:

- full-depth involute with $\phi = 14.5°$: $a = \dfrac{1}{P_d}$ and $b = \dfrac{1.157}{P_d}$;
- full-depth involute with $\phi = 20°$: $a = \dfrac{1}{P_d}$ and $b = \dfrac{1.157}{P_d}$;
- stub involute with $\phi = 20°$: $a = \dfrac{8}{10P_d}$ and $b = \dfrac{1}{P_d}$.

The gear interference is defined as the contact of parts of tooth profiles that are not conjugate. This can be the case when segments of teeth that are not involute will interfere during gear meshing.

The maximum possible addendum circle radius without interference for external gearing is

$$r_{a(max)} = \sqrt{r_b^2 + c^2 \sin^2 \phi} \quad \text{or}$$

$$r_{a(max)p} = \sqrt{r_{bp}^2 + c^2 \sin^2 \phi} \quad \text{and} \quad r_{a(max)g} = \sqrt{r_{bg}^2 + c^2 \sin^2 \phi}, \tag{6.6}$$

where $r_{bg} = r_g \cos\phi$ and $r_{bp} = r_p \cos\phi$ are the base circle radii of the gear and pinion, respectively. The pitch base, p_b, is given by

$$p_b = \frac{\pi\, d_b}{N} \quad \text{or} \quad p_b = \frac{\pi\, d_{bp}}{N_p} = \frac{\pi\, d_{bg}}{N_g}. \tag{6.7}$$

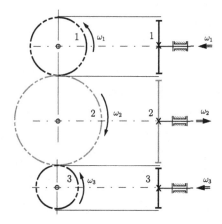

Figure 6.4 Simple gear train.

A link of two or more contacting gears forms a gear train. Fig. 6.4 represents a gear train formed by three gears connected in series. The ratio of the gear train is

$$i_{13} = \frac{\omega_1}{\omega_3} = \frac{\omega_1}{\omega_2}\frac{\omega_2}{\omega_3} = \left(-\frac{N_2}{N_1}\right)\left(-\frac{N_3}{N_2}\right) = \frac{N_3}{N_1}. \tag{6.8}$$

The middle gear 2 is an idler and does not affect the overall ratio.

The gears of a gear train rotate about their own axes. An epicyclic (or planetary) gear train has gear axes that rotate also with respect to a reference frame. When a circle (planet gear) rolls without slipping on the circumference of another circle (directing circle or sun gear), any point of the circle describes an epicycloid. The parametric equation of the epicycloids is given by

$$x = (b+a)\cos\theta - b\cos\frac{(b+a)\theta}{b},$$

$$y = (b+a)\sin\theta - b\sin\frac{(b+a)\theta}{b}.$$

The MATLAB program for the epicycloid shown in Fig. 6.5 is given by:

```
b = 0.1;
a = 0.1;
theta = 0:pi/100:2*pi;
% circle of radius a (sun)
xa = a*cos(theta);
```

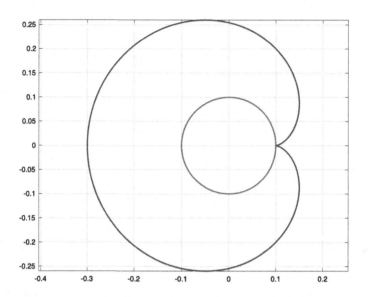

Figure 6.5 Epicycloid curve.

```
ya = a*sin(theta);
% parametric equation of epicycloid
xi = (b + a)*cos(theta) - b*cos((b + a)*theta./b);
yi = (b + a)*sin(theta) - b*sin((b + a)*theta./b);
figure
plot(xa,ya,'-r', xi,yi,'-b','linewidth',2)
grid on
axis equal
```

Example 6.1. A planetary gear train is shown in Fig. 6.6. The system consists of a directing/sun gear 1 with an angular speed of 75 rpm, two generating/planet gears 2 and 2′ with $N_2 = 20$ teeth, and a fixed ring gear 4 with $N_4 = 70$ teeth. The module of the gears is $m = 2$ mm. If arm 3 drives a machine, determine its angular speed.

Solution

The planet gear 2′ is a passive element. This planet gear can be eliminated, and the number of degrees of freedom of the mechanism remains the same. To calculate the number of degrees of freedom for the planetary gear train, the passive element must be eliminated as shown in Fig. 6.7. The number of degrees of freedom of the epicyclic gear train is calculated by

$$M = 3n - 2c_5 - c_4,$$

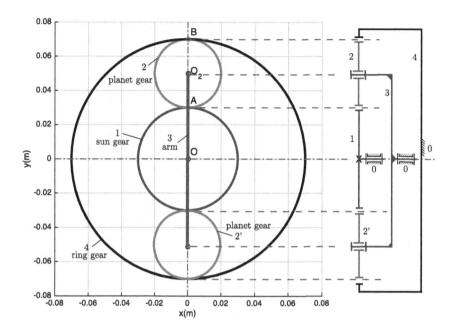

Figure 6.6 Planetary gear train.

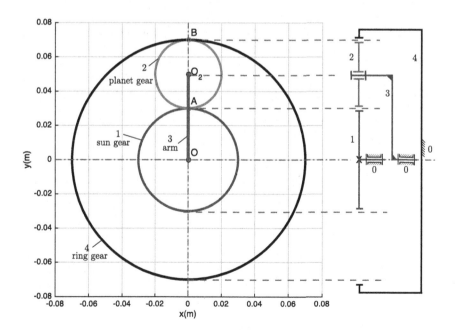

Figure 6.7 Planetary gear train without the passive planet gear.

where n, c_5, and c_4 represent the number of moving links, and the number of one and two degree of freedom joints, respectively. There are three moving links (gear 1, gear 2, and arm 3) connected by joints, $n = 3$. There is a single one degree of freedom joint (revolute joint) between the gear 1 and fixed frame 0 located at O. There is a revolute joint between gear 2 and arm 3 located at O_2. There is a gear joint (two degree of freedom joint) between gear 1 and gear 2 at A, and between gear 2 and gear 4 at B. The ring gear 4 is fixed to frame 0. There is a revolute joint between the fixed frame 0 and planet arm 3 at O. The number of degrees of freedom of the epicyclic gear train is $M = 3n - 2c_4 - c_4 = 3(3) - 2(3) - 2 = 1$.

The gears will mesh at their pitch circles. The pitch circle radii of the planet and ring gear are $r_2 = mN_2/2 = 2(20)/2 = 20$ mm and $r_4 = mN_4/2 = 2(70)/2 = 70$ mm. The circle radius of the sun gear is $r_1 = r_4 - 2r_2 = 30$ mm.

The position vectors $\mathbf{r}_A$, $\mathbf{r}_{O_2}$, and $\mathbf{r}_B$ are

$$\mathbf{r}_A = x_A \mathbf{1} + y_A \mathbf{J} = r_1 \mathbf{J},$$
$$\mathbf{r}_{O_2} = x_{O_2} \mathbf{1} + y_{O_2} \mathbf{J} = (r_1 + r_2) \mathbf{J},$$
$$\mathbf{r}_B = x_B \mathbf{1} + y_B \mathbf{J} = r_4 \mathbf{J}.$$

The MATLAB commands for position and graphic analysis are:

```
m = 2*10^-3; % (m)
N2 = 20; % (teeth)
N4 = 70; % (teeth)
% pitch radii
r2 = m*N2/2; % (m)
r4 = m*N4/2; % (m)
r1 = r4 - 2*r2;
% origin of the reference frame
% center of gear 1: 01=0
r01_ = [0 0 0];
% center of gear 4: 04
r04_ = r01_;
% center of gear 2: 02
r02_ = [0 r1+r2 0];

rA_ = [0 r1 0];
rB_ = [0 r4 0];
```

```
figure(1)
af = .08; axis([-af af -af af])
axis equal
axis square
grid on
hold on
xlabel('x(m)'), ylabel('y(m)'), zlabel('z(m)')
t = linspace(0,2*pi);
% plot gear 1 of radius r1 at 01
plot(r01_(1)+r1*cos(t),r01_(2)+r1*sin(t),...
 'color','b','LineWidth',3)
% plot gear 2 of radius r2 at 02
plot(r02_(1)+r2*cos(t),r02_(2)+r2*sin(t),...
 'color','r','LineWidth',3)
plot(r04_(1)+r4*cos(t),r04_(2)+r4*sin(t),...
 'color','k','LineWidth',3)
text(r01_(1),r01_(2)+0.004,' 0','fontsize',12)
text(r02_(1),r02_(2),' 0_2','fontsize',12)
text(rA_(1),rA_(2)+0.004,' A','fontsize',12)
text(rB_(1),rB_(2)+0.004,' B','fontsize',12)
plot(r01_(1),r01_(2),'o','color','k')
plot(r02_(1),r02_(2),'o','color','k')
plot(rA_(1),rA_(2),'o','color','r')
plot(rB_(1),rB_(2),'o','color','r')
line([r01_(1),r02_(1)],[r01_(2),r02_(2)],...
    'LineStyle','-','LineWidth',4)
line([-af,af],[0,0],...
    'LineStyle','-.','LineWidth',1.5)
line([0,0],[-af,af],...
    'LineStyle','-.','LineWidth',1.5)
```

The velocity of point A_1 on the sun gear 1 is

$$\mathbf{v}_{A_1} = \mathbf{v}_O + \boldsymbol{\omega}_1 \times \mathbf{r}_A = 0 + \begin{vmatrix} \mathbf{1} & \mathbf{J} & \mathbf{k} \\ 0 & 0 & \omega_1 \\ 0 & r_1 & 0 \end{vmatrix} = -\omega_1\, r_1\, \mathbf{1}.$$

The velocity of point A_2 on the planet gear 2 is

$$\mathbf{v}_{A_2} \;=\; \mathbf{v}_{B_2} + \boldsymbol{\omega}_2 \times \mathbf{r}_{BA} = 0 + \begin{vmatrix} \mathbf{1} & \mathbf{J} & \mathbf{k} \\ 0 & 0 & \omega_2 \\ 0 & -2r_2 & 0 \end{vmatrix} = 2\,\omega_2\,r_2\,\mathbf{1}.$$

The pitch line velocities at A are equal, $\mathbf{v}_{A_1} = \mathbf{v}_{A_2}$. The planet arm angular velocity is calculated as

$$\omega_2 = -\omega_1\, r_1/(2\,r_2).$$

The velocity of center O_2 on link 2 is

$$\mathbf{v}_{O_2} \;=\; \mathbf{v}_{B_2} + \boldsymbol{\omega}_2 \times \mathbf{r}_{BO_2} = 0 + \begin{vmatrix} \mathbf{1} & \mathbf{J} & \mathbf{k} \\ 0 & 0 & \omega_2 \\ 0 & -r_2 & 0 \end{vmatrix} = \omega_2\,r_2\,\mathbf{1} = -0.5\,\omega_1\,r_1\,\mathbf{1}.$$

The velocity of O_2 on link 3 is

$$\mathbf{v}_{O_2} \;=\; \mathbf{v}_O + \boldsymbol{\omega}_3 \times \mathbf{r}_{O_2} = 0 + \begin{vmatrix} \mathbf{1} & \mathbf{J} & \mathbf{k} \\ 0 & 0 & \omega_3 \\ 0 & r_1 + r_2 & 0 \end{vmatrix} = -\omega_3\,(r_1 + r_2)\,\mathbf{1}.$$

From the previous equations the angular velocity of arm 3 is

$$\omega_3 = 0.5\,\omega_1\, r_1/(r_1 + r_2) = 18.850 \quad \text{rad/s}.$$

The MATLAB code for the angular velocity of the planet arm is:

```
n1 = 75;              % (rpm)
omega1 = n1*pi/30;    % (rad/s)
omega1_ = [0 0 omega1];
v0_ = [0 0 0];
vA1_ = v0_ + cross(omega1_,rA_);

syms omega2
vB2_ = [0 0 0];
omega2_ = [0 0 omega2];
vA2_ = vB2_ + cross(omega2_,rA_-rB_);

eqvA_ = vA1_-vA2_;
eqvA = eqvA_(1);

omega2 = eval(solve(eqvA));
```

```
omega2_ = [0 0 omega2];

syms omega3
omega3_ = [0 0 omega3];
v022_ = vB2_ + cross(omega2_,r02_-rB_);
v023_ = v0_ + cross(omega3_,r02_);
eqv02_ = v022_-v023_;
eqv02 = eqv02_(1);

omega3  = eval(solve(eqv02));
omega3_ = [0 0 omega3];
```

The results are:

```
r1 =  0.030 (m)
r2 =  0.020 (m)
r4 =  0.070 (m)
omega1= 7.854(rad/s)=75.000(rpm)

omega2=-5.890(rad/s)=-56.250(rpm)
omega3= 2.356(rad/s)=22.500(rpm)
```

Example 6.2. The gears of the planetary mechanism shown in Fig. 6.8 mesh at their pitch circles. The pitch radius of gear 1 (sun) is $r_1 = 4$ in and the pitch radius of gear 2 (planet) is $r_2 = 4$ in. Planet gear 2' with the pitch radius of $r_{2'} = 8$ in is fixed – as planet gear 2 – on the same shaft. At a given instant the sun gear 1 has angular velocity $\omega_1 = -3$ rad/s (clockwise) and angular acceleration $\alpha_1 = -7$ rad/s^2 (clockwise) and the outer ring gear 3 has angular velocity $\omega_3 = 2$ rad/s (counterclockwise) and angular acceleration $\alpha_3 = 5$ rad/s^2 (counterclockwise). Find the angular acceleration of planet gear 2 at this instant.

Solution

The input data for the planetary gear train are introduced in MATLAB with the commands:

```
syms r1 r2 r2p
syms omega1 omega2 omega3 alpha1 alpha2 alpha3
% simbolical values
lists = {r1, r2, r2p, omega1, omega3, alpha1, alpha3};
% numerical values
listn = {4, 4, 8, -3, 2, -7, 5};
```

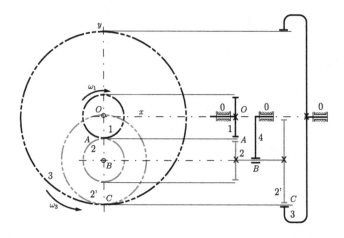

Figure 6.8 Planetary gear train.

and the position vectors of points A, B, and C are

```
rA_ = [0 -r1 0];
rB_ = [0 -r1-r2 0];
rC_ = [0 -r1-r2-r2p 0];
```

The velocity at A is

$$\mathbf{v}_A = \mathbf{v}_O + \boldsymbol{\omega}_1 \times \mathbf{r}_A = \boldsymbol{\omega}_1 \times \mathbf{r}_A.$$

The velocity of C on the planet gear $2'$ is calculated as

$$\mathbf{v}_{C_2} = \mathbf{v}_A + \boldsymbol{\omega}_2 \times \mathbf{r}_{AC}.$$

The velocity of C on the ring gear 3 is

$$\mathbf{v}_{C_3} = \mathbf{v}_O + \boldsymbol{\omega}_3 \times \mathbf{r}_3.$$

The angular velocity $\boldsymbol{\omega}_2$ is calculated from $\mathbf{v}_{C_2} = \mathbf{v}_{C_3}$, or in MATLAB:

```
v0_ = [0 0 0];
vA_ = v0_ + cross(omega1_,rA_);
% velocity of C on planet gear 2p
vC2_ = vA_ + cross(omega2_,rC_-rA_);
% velocity of C on ring gear 3
vC3_ = v0_ + cross(omega3_,rC_);
```

```
eqvC_ = vC2_-vC3_;
eqvC  = eqvC_(1);
omega2 = solve(eqvC,omega2);
omega2n = eval(subs(omega2,lists, listn));

% omega2 = -(omega1*r1 - omega3*(r1 + r2 + r2p))/(r2 + r2p)
% omega2 =  3.667 (rad/s)
```

The angular acceleration ω_2 is calculated from $\mathbf{a}_{C_2} = \mathbf{a}_{C_3}$, or in MAT-LAB:

```
alpha1_ = [0 0 alpha1];
alpha2_ = [0 0 alpha2];
alpha3_ = [0 0 alpha3];
a0_ = [0 0 0];
aA_ = a0_ + cross(alpha1_,rA_)...
    +cross(omega1_,cross(omega1_,rA_));
aC2_ = aA_ + cross(alpha2_,rC_-rA_)...
    +cross(omega2_,cross(omega2_,rC_-rA_));
aC3_ = a0_ + cross(alpha3_,rC_)...
    +cross(omega3_,cross(omega3_,rC_));
eqaC_ = aC2_-aC3_;
eqaC  = eqaC_(1);
alpha2 = solve(eqaC,alpha2);
alpha2n = eval(subs(alpha2,lists, listn));
% alpha2 = -(alpha1*r1 - alpha3*(r1 + r2 + r2p))/(r2 + r2p)
% alpha2 =  9.000 (rad/s^2)
```

6.2. Gear forces

For the next gear force calculation friction is assumed negligible, and the gears are considered meshing along their pitch circles. The force between two contacting frictionless gears is calculated – as shown in Fig. 6.9 – at the pitch point P. Its tangential component F_t relates with the power transmitted while its radial component F_r tends to separate the gears. The two components are related by the equation

$$F_r = F_t \tan\phi, \tag{6.9}$$

where the pressure angle is ϕ. In Fig. 6.9, gear 1 is the driver and gear 2 is the driven gear. The pitch line velocity, V, is equal to

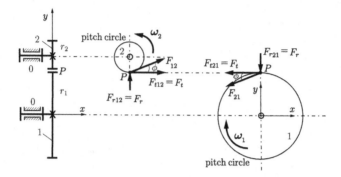

Figure 6.9 Gear forces at the pitch point. From Dan B. Marghitu, Kinematic Chains and Machine Elements Design, Elsevier, 2005.

$$V(\text{ft/min}) = \pi \, d(\text{in}) \, n(\text{rpm})/12,$$
$$V(\text{m/s}) = \pi \, d(\text{mm}) \, n(\text{rpm})/60\,000, \qquad (6.10)$$

where the pitch diameter d is expressed in inches or mm, and n is the angular speed in rpm.

The transmitted power and torque are calculated with

$$H(\text{hp}) = \frac{1}{33\,000} F_t V,$$
$$M_t \, (\text{lb in}) = \frac{1}{63\,000} H/n, \qquad (6.11)$$

where F_t is expressed in pounds, V are in feet per minute, H in horsepower, and n in rpm. The transmitted power and torque can also be calculated with

$$H(\text{W}) = F_t V,$$
$$M_t \, (\text{N m}) = 9549 \, H/n, \qquad (6.12)$$

where F_t is expressed in newtons, V in meters per seconds, H in kW and n in rpm.

Example 6.3. The two stage gear train shown in Fig. 6.10 has an input driver gear 1, with $N_1 = 21$ teeth. The driver gear is coupled with a motor having power $H = 1$ kW and speed $n = n_1 = 1200$ rpm. Gear 2 having $N_2 = 63$ teeth and gear 2' having $N_{2'} = 21$ teeth are rigidly connected on the same countershaft which rotates freely in bearings A and B. The output gear 3 having $N_3 = 63$ teeth is rigidly fixed to another shaft. Each stage has a diametral pitch of $P_d = 5$, and a pressure angle of $\phi = 25°$.

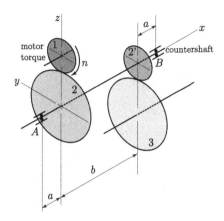

Figure 6.10 Two stage gear train.

The distance between the gears is $b = 300$ mm, and the distance between the gears and bearings is $a = 100$ mm. Neglecting the friction loss in the gears and bearings, determine the applied forces to countershaft bearings A and B.

Solution
The pitch radius of gear i is

$$r_i = m \, N_i / 2,$$

where $m = 25.4/Pd$ is the module. In MATLAB the gear pitch radii are calculated with:

```
% diametral pitch
Pd = 5; % (teeth/inch)
% module
m = 25.4/Pd; % (mm)
% number of teeth
N1 = 21; % (teth)
N2 = 63; % (teth)
N2p= 21; % (teth)
N3 = 63; % (teth)
% pitch radii
r1 = m*N1/2;  % (mm)
r2 = m*N2/2;  % (mm)
r2p= m*N2p/2; % (mm)
```

```
r3 = m*N3/2;  % (mm)
r1  = r1*10^(-3); % (m)
r2  = r2*10^(-3); % (m)
r2p = r2p*10^(-3);% (m)
r3  = r3*10^(-3); % (m)


% r1  =  0.053 (m)
% r2  =  0.160 (m)
% r2p =  0.053 (m)
% r3  =  0.160 (m)
```

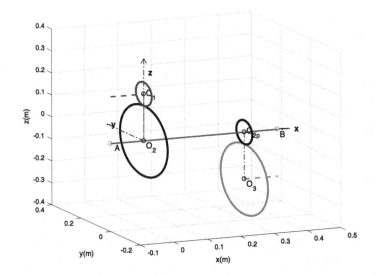

Figure 6.11 MATLAB figure for two stage gear train.

The MATLAB figure of the gear, shown in Fig. 6.11, can be obtained with:

```
a = 0.1; % (m)
b = 0.3; % (m)

% origin of the reference frame
% center of gear 2: O2
rO2_ = [0 0 0];
% center of gear 1: O1
rO1_ = [0 0 r1+r2];
% center of gear 2p: O2p
rO2p_ = [b 0 0];
```

```
% center of gear 3: O3
r03_ = [b 0 -(r2p+r3)];

% F12_ acts at P
rP_ = [0 0 r2];
% F32_ acts at Q
rQ_ = [b 0 -r2p];
% bearing at A
rA_ = [-a 0 0];
% bearing at B
rB_ = [a+b 0 0];

figure(1)
af = .25;
axis([-af af -af af -af af])
grid on
hold on
axis auto
xlabel('x(m)'), ylabel('y(m)'), zlabel('z(m)')
t = linspace(0,2*pi);
% plot gear 1 of radius r1 at O1
plot3(0*t,r01_(2)+r1*cos(t),r01_(3)+r1*sin(t),...
 'color','b','LineWidth',3)
% plot gear 2 of radius r2 at O2
plot3(0*t,r2*cos(t),r2*sin(t),...
 'color','k','LineWidth',3)
% plot gear 2p of radius r2p at O2p
plot3(r02p_(1)+0*t,r02p_(2)+r2p*cos(t),r02p_(3)+r2p*sin(t),...
 'color','k','LineWidth',3)
% plot gear 3 of radius r3 at O3
plot3(r03_(1)+0*t,r03_(2)+r3*cos(t),r03_(3)+r3*sin(t),...
 'color','r','LineWidth',3)

text(r01_(1),r01_(2),r01_(3),' O_1','fontsize',12)
text(r02_(1),r02_(2),r02_(3),' O_2','fontsize',12)
text(r02p_(1),r02p_(2),r02p_(3),' O_{2p}','fontsize',12)
text(r03_(1),r03_(2),r03_(3),' O_3','fontsize',12)
text(rA_(1),rA_(2),rA_(3),'   A','fontsize',12)
text(rB_(1),rB_(2),rB_(3),' B','fontsize',12)

plot3(r01_(1),r01_(2),r01_(3),'o','color','k')
plot3(r02_(1),r02_(2),r02_(3),'o','color','k')
```

```
plot3(r02p_(1),r02p_(2),r02p_(3),'o','color','k')
plot3(r03_(1),r03_(2),r03_(3),'o','color','k')

plot3(rA_(1),rA_(2),rA_(3),'o','color','r')
plot3(rB_(1),rB_(2),rB_(3),'o','color','r')

line([r01_(1),r02_(1)],[r01_(2),r02_(2)],[r01_(3),r02_(3)],...
    'LineStyle','-.','LineWidth',1.5)
line([rA_(1),rB_(1)],[rA_(2),rB_(2)],[rA_(3),rB_(3)],...
    'LineStyle','-','LineWidth',2)
line([r02p_(1),r03_(1)],[r02p_(2),r03_(2)],[r02p_(3),r03_(3)],...
    'LineStyle','-.','LineWidth',1.5)

line([r01_(1)-a,r01_(1)],[r01_(2),r01_(2)],[r01_(3),r01_(3)],...
    'LineStyle','--','LineWidth',2,'Color','b')
line([r03_(1),r03_(1)+a],[r03_(2),r03_(2)],[r03_(3),r03_(3)],...
    'LineStyle','--','LineWidth',2,'Color','r')

% cartesian axes
quiver3(0,0,0,1.75*af,0,0,1,'Color','k','LineWidth',1,'LineStyle','-.')
quiver3(0,0,0,0,af,0,1,'Color','k','LineWidth',1,'LineStyle','-.')
quiver3(0,0,0,0,0,1.5*af,1,'Color','k','LineWidth',1,'LineStyle','-.')
text(1.75*af,0,0,'  x','fontsize',12,'fontweight','b')
text(0,af,0,'  y','fontsize',12,'fontweight','b')
text(0,0,1.25*af,'  z','fontsize',12,'fontweight','b')
view(-24,14)
```

The angular speed of countershaft 2 is

$$n_2 = -n_1 \frac{N_1}{N_2},$$

and for the third gear and shaft

$$n_3 = -n_2 \frac{N_{2'}}{N_3},$$

or in MATLAB:

```
n1 = 1200; % (rpm)
n2 = -n1*N1/N2; % (rpm)
n3 = -n2*N2p/N3; % (rpm)
% n2 = -400.000 (rpm)
% n3 =  133.333 (rpm)
```

The torque M_1 in driver shaft 1 is

$$M_1 = \frac{9549\,H_1}{n_1},$$

where $H_1 = H$ is the power of the motor. The transmitted torques for each shaft are calculated with:

```
% torque on the shaft 1
M1 = 9549*H/n1; % (N m)
% torque on the shaft 2
M2 = 9549*H/n2; % (N m)
% torque on the shaft 3
M3 = 9549*H/n3; % (N m)
% M1 =   7.957 (N m)
% M2 = -23.872 (N m)
% M3 = 71.617 (N m)
```

All the gear forces are calculated at the pitch points P and Q. The tangential force of gear 2 on motor pinion 1 at P is

$$F_{t21} = \frac{M_1}{r_1},$$

and the motor pinion radial force is

$$F_{r21} = F_{t21} \tan \phi.$$

The forces of the countershaft pinion $2'$ on gear 3 at Q are

$$F_{t2'3} = \frac{M_3}{r_3} \quad \text{and} \quad F_{r2'3} = F_{t23} \tan \phi.$$

The MATLAB results are:

```
Ft21 = M1/r1;
Fr21 = Ft21*tan(phi);
Ft2p3 = M3/r3;
Fr2p3 = Ft23*tan(phi);
% Ft21 = 149.184 (N)
% Fr21 = 69.566 (N)
%
% Ft2p3 = 447.553 (N)
% Fr2p3 = 208.698 (N)
```

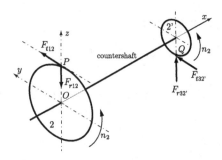

Figure 6.12 Gear forces on the countershaft.

The vector force of the motor pinion 1 on gear 2 at P, as shown in Fig. 6.12, is

$$\mathbf{F}_{12} = F_{r21}\mathbf{J} - F_{t21}\mathbf{k},$$

and the vector force of gear 3 on gear 2' at Q is

$$\mathbf{F}_{32'} = F_{r2'3}\mathbf{J} + F_{t2'3}\mathbf{k},$$

or in MATLAB:

```
F12_ = [0 Ft21 -Fr21];
F32p_ = [0 Ft2p3  Fr2p3];
% F12_ = [ 0 149.184 -69.566] (N)
% F32p_ = [ 0 447.553 208.698] (N)
```

The radial bearing forces at A and B are

$$\mathbf{F}_A = F_{Ay}\mathbf{J} + F_{Az}\mathbf{k},$$
$$\mathbf{F}_B = F_{By}\mathbf{J} + F_{Bz}\mathbf{k}.$$

The countershaft free-body diagram is shown in Fig. 6.13. The reaction force $\mathbf{F}_A$ is calculated from

$$\sum \mathbf{M}_B = \mathbf{r}_{BP} \times \mathbf{F}_{12} + \mathbf{r}_{BQ} \times \mathbf{F}_{32'} + \mathbf{r}_{BA} \times \mathbf{F}_A = 0.$$

The MATLAB commands for the reaction at A are:

```
syms FAx FAy FAz FBx FBy FBz
FA_ = [FAx, FAy, FAz];
FB_ = [FBx, FBy, FBz];
```

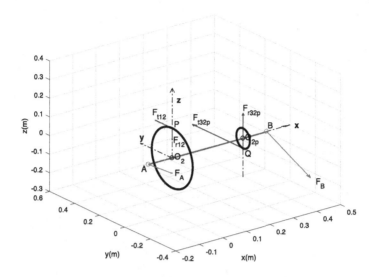

Figure 6.13 Free-body diagram of the countershaft.

```
% sum MB_ = rBP_ x F12_ + rBA_ x FA_ + rBQ_ x F32p_ = 0_ =>
MB_ = cross(rP_-rB_,F12_) + cross(rA_-rB_,FA_) + cross(rQ_-rB_,F32p_);
solFA = solve(MB_(2),MB_(3));
FAys=eval(solFA.FAy);
FAzs=eval(solFA.FAz);
FA_ = [0,FAys,FAzs];
% FA_ = [ 0 -208.858 13.913] (N)
```

The radial force at B is determined from the sum of all forces that act on the countershaft

$$\sum \mathbf{F} = \mathbf{F}_{12} + \mathbf{F}_A + \mathbf{F}_B + \mathbf{F}_{32'} = 0,$$

or in MATLAB:

```
% FB_ + FA_ + F12_ + F32p_ = 0
FB_ = -FA_-F12_-F32p_;
% FB_ = [0 -387.880 -153.045] (N)
```

The MATLAB instructions for sketching the forces on countershaft are:

```
figure(2)
% scale factor for forces
sf = 1000;
```

```
af = .25;
axis([-af af -af af -af af])
grid on
hold on
axis auto
xlabel('x(m)'), ylabel('y(m)'), zlabel('z(m)')
t = linspace(0,2*pi);
% plot gear 2 of radius r2 at O2
plot3(0*t,r2*cos(t),r2*sin(t),...
 'color','k','LineWidth',3)
% plot gear 2p of radius r2p at O2p
plot3(rO2p_(1)+0*t,rO2p_(2)+r2p*cos(t),rO2p_(3)+r2p*sin(t),...
 'color','k','LineWidth',3)

text(rO2_(1),rO2_(2),rO2_(3),' O_2','fontsize',12)
text(rO2p_(1),rO2p_(2),rO2p_(3),' O_{2p}','fontsize',12)
text(rA_(1),rA_(2),rA_(3),'   A','fontsize',12)
text(rB_(1),rB_(2),rB_(3),' B','fontsize',12)

plot3(rO2_(1),rO2_(2),rO2_(3),'o','color','k')
plot3(rO2p_(1),rO2p_(2),rO2p_(3),'o','color','k')

plot3(rA_(1),rA_(2),rA_(3),'o','color','r')
plot3(rB_(1),rB_(2),rB_(3),'o','color','r')

line([rO1_(1),rO2_(1)],[rO1_(2),rO2_(2)],[rO1_(3),rO2_(3)],...
    'LineStyle','-.','LineWidth',1.5)
line([rA_(1),rB_(1)],[rA_(2),rB_(2)],[rA_(3),rB_(3)],...
    'LineStyle','-','LineWidth',2)
line([rO2p_(1),rO3_(1)],[rO2p_(2),rO3_(2)],[rO2p_(3),rO3_(3)],...
    'LineStyle','-.','LineWidth',1.5)

% plot F12_ at P
quiver3(...
 rP_(1),rP_(2),rP_(3),...
 0,F12_(2)/sf,0,'color','b','LineWidth',1.3);
quiver3(...
 rP_(1),rP_(2),rP_(3),...
  0,0,F12_(3)/sf,'color','b','LineWidth',1.3);
```

```
text(rP_(1),rP_(2),rP_(3),'  P','fontsize',12)
text(rP_(1),rP_(2)+F12_(2)/sf,rP_(3),'F_{t12}','fontsize',12)
text(rP_(1),rP_(2),rP_(3)+F12_(3)/sf,'F_{r12}','fontsize',12)

% plot F32p_ at Q
quiver3(...
  rQ_(1),rQ_(2),rQ_(3),...
  0,F32p_(2)/sf,0,'color','b','LineWidth',1.3);
quiver3(...
  rQ_(1),rQ_(2),rQ_(3),...
  0,0,F32p_(3)/sf,'color','b','LineWidth',1.3);

text(rQ_(1),rQ_(2),rQ_(3),'  Q','fontsize',12)
text(rQ_(1),rQ_(2)+F32p_(2)/sf,rQ_(3),'F_{t32p}','fontsize',12)
text(rQ_(1),rQ_(2),rQ_(3)+F32p_(3)/sf,'F_{r32p}','fontsize',12)

% plot FA_ at A
quiver3(...
  rA_(1),rA_(2),rA_(3),...
  FA_(1)/sf,FA_(2)/sf,FA_(3)/sf,'color','r','LineWidth',1.3);
text(rA_(1),rA_(2)+FA_(2)/sf,rA_(3)+FA_(3)/sf,'F_{A}','fontsize',12)

% plot FB_ at B
quiver3(...
  rB_(1),rB_(2),rB_(3),...
  FB_(1)/sf,FB_(2)/sf,FB_(3)/sf,'color','r','LineWidth',1.3);
text(rB_(1),rB_(2)+FB_(2)/sf,rB_(3)+FB_(3)/sf,'F_{B}','fontsize',12)

% cartesian axes
quiver3(0,0,0,2*af,0,0,1,'Color','k','LineWidth',1,'LineStyle','-.')
quiver3(0,0,0,0,af,0,1,'Color','k','LineWidth',1,'LineStyle','-.')
quiver3(0,0,0,0,0,1.5*af,1,'Color','k','LineWidth',1,'LineStyle','-.')

text(2*af,0,0,'  x','fontsize',12,'fontweight','b')
text(0,af,0,'  y','fontsize',12,'fontweight','b')
text(0,0,1.25*af,'  z','fontsize',12,'fontweight','b')
```

Example 6.4. The gear train, shown in Fig. 6.14A, transmits the power of $H = 1$ kW through an electric motor coupled to gear 1 having the number of teeth $N_1 = 20$ and angular speed $n = n_1 = 1000$ rpm. Gear 2 having $N_2 = 64$ teeth and gear 2' having $N_{2'} = 20$ teeth are rigidly connected on the same countershaft which rotates freely in bearings A and B. Gear 3 (the

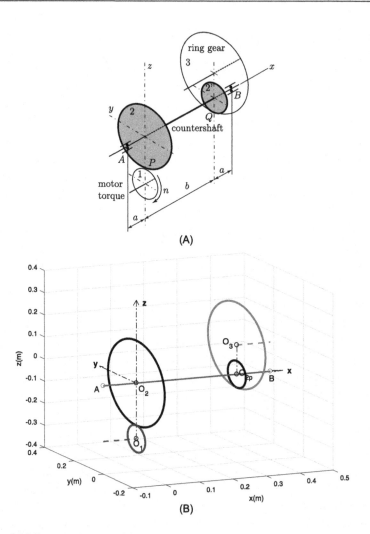

Figure 6.14 Two stage gear train.

output ring) is rigidly fixed to a shaft coupled to the driven machine and has $N_3 = 64$ teeth. The shafts of the gears are parallel. The module for each stage is $m = 6$ mm and the pressure angle is $\phi = 20°$. The distance between the bearings and gears is $a = 100$ mm and the distance between the gears is $b = 300$ mm. Find the countershaft bearings reaction forces at A and B. The friction loss is neglected.

Solution

The pitch radii of the gears are:

```
% no. of teeth for each gear
```

```
N1 = 20; % (teeth)
N2 = 64; % (teeth)
N2p= 20; % (teeth)
N3 = 64; % (teeth)
% module
m = 6; % (mm)
% pressure angle
phi = 20*pi/180; % (rad)
% pitch radii
r1 = m*N1/2;   % (mm)
r2 = m*N2/2;   % (mm)
r2p= m*N2p/2;  % (mm)
r3 = m*N3/2;   % (mm)
r1 = r1*10^(-3);   % (m)
r2 = r2*10^(-3);   % (m)
r2p= r2p*10^(-3);  % (m)
r3 = r3*10^(-3);   % (m)

% r1 =   0.060 (m)
% r2 =   0.192 (m)
% r2p =  0.060 (m)
% r3 =   0.192 (m)
```

The MATLAB figure of the gear train is shown in Fig. 6.14B. The angular speeds of the gears are

```
n1 = 1000; % (rpm)
n2 = -n1*N1/N2; % (rpm)
n3 =  n2*N2p/N3; % (rpm)
% n1 = 1000.000 (rpm)
% n2 = -312.500 (rpm)
% n3 = -97.656 (rpm)
```

The transmitted torques are:

```
% power on gear 1
H = 1; % (kW)
% torque on the shaft 1
M1 = 9549*H/n1; % (N m)
% torque on the shaft 2
M2 = 9549*H/n2; % (N m)
```

```
% torque on the shaft 3
M3 = 9549*H/n3; % (N m)
% M1 =   9.549 (N m)
% M2 = -30.557 (N m)
% M3 = -97.782 (N m)
```

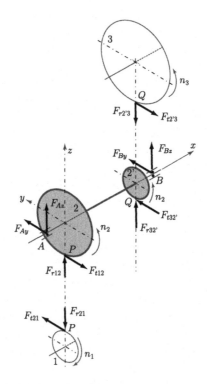

Figure 6.15 Force analysis for the gear train.

The free-body diagrams of the gears are shown in Fig. 6.15. The tangential and radial forces are calculated with:

```
% tangential force of gear 2 on motor pinion 1 at P
Ft21 = M1/r1;
% radial force of gear 2 on 1 at P
Fr21 = Ft21*tan(phi);
% gear force of  2 on 1 at P
F21_ = [0 Ft21 -Fr21];
% force of gear 1 on 2 at P
F12_ = -F21_;
```

```
% tangential force of gear 3 on 2 at Q
Ft32p = abs(M2/r2p);
% radial force of gear 3 on 2p at Q
Fr32p = Ft32p*tan(phi);
% force of gear 3 on 2p at Q
F32p_ = [0 Ft32p Fr32p];
% force of gear 2p on 3 at Q
F2p3_ = -F32p_;
% |Ft21| = 159.150 (N)
% |Fr21| = 57.926 (N)
% F21_ = [ 0 159.150 -57.926] (N)
% F12_ = [ 0 -159.150 57.926] (N)
%
% |Ft32p| = 509.280 (N)
% |Fr32p| = 185.363 (N)
% F32p_ = [ 0 509.280 185.363] (N)
% F2p3_ = [ 0 -509.280 -185.363] (N)
```

The radial bearing forces at A and B are calculated with:

```
syms FAx FAy FAz FBx FBy FBz
FA_ = [FAx, FAy, FAz];
FB_ = [FBx, FBy, FBz];
% sum MB_ = rBP_ x F12_ + rBA_ x FA_ + rBQ_ x F32_ = 0_ =>
MB_ = cross(rP_-rB_,F12_) + cross(rA_-rB_,FA_) + cross(rQ_-rB_,F32p_);
solFA = solve(MB_(2),MB_(3));
FAys=eval(solFA.FAy);
FAzs=eval(solFA.FAz);
FA_ = [0,FAys,FAzs];
FAr =sqrt(FA_(2)^2+FA_(3)^2);
% FB_ + FA_ + F12_ + F32p_ = 0
FB_ = -FA_-F12_-F32p_;
FBr =sqrt(FB_(1)^2+FB_(3)^2);
% MBx = 0.0
% MBy = 0.5*FAz + 41.7066
% MBz = 12.732 - 0.5*FAy
%
% FA_ = [ 0 25.464 -83.413] (N)
% radial force FAr = 87.213 (N)
%
% FB_ = [0 -375.594 -159.875] (N)
% radial force FBr = 159.875 (N)
```

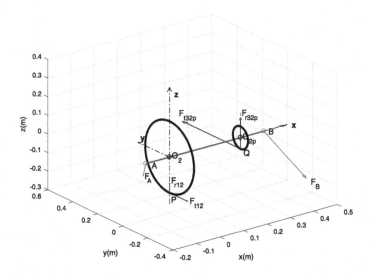

Figure 6.16 Forces acting on the countershaft.

The forces on the countershaft are shown in Fig. 6.16.

6.3. Gear design

The forces on a gear tooth are shown in Fig. 6.17. The procedure proposed in [7] is followed. The tangential component of the force, F_t, producing the bending stress, σ, at the base of the tooth given by

$$\sigma = \frac{M}{I}c = \frac{M\ (t/2)}{B\ t^3/12} = \frac{6M}{Bt^2} = \frac{6F_t h}{Bt^2},$$

is

$$F_t = \sigma B\frac{t^2}{6h} = \sigma Bp\frac{t^2}{6hp}, \qquad (6.13)$$

where h is the distance from the application point of the force to the base of the tooth, t is the tooth thickness, B is the width of the face, and the bending moment is $M = F_t h$. The face width is restricted by $B = kp$, where $k \leq 4$, and p is the circular pitch. The dimensionless form factor is defined by the equation

$$y = \frac{t^2}{6hp}. \qquad (6.14)$$

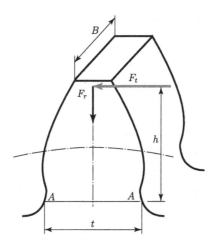

Figure 6.17 Gear tooth.

For 20° full-depth involute teeth, the form factor γ for use in Lewis strength equation is given by [7]

$$\gamma = \begin{cases} 0.078 & \text{for } N = 12 \text{ teeth,} \\ 0.083 & \text{for } N = 13 \text{ teeth,} \\ 0.088 & \text{for } N = 14 \text{ teeth,} \\ 0.092 & \text{for } N = 15 \text{ teeth,} \\ 0.094 & \text{for } N = 16 \text{ teeth,} \\ 0.096 & \text{for } N = 17 \text{ teeth,} \\ 0.098 & \text{for } N = 18 \text{ teeth,} \\ 0.100 & \text{for } N = 19 \text{ teeth,} \\ 0.102 & \text{for } N = 20 \text{ teeth,} \\ 0.104 & \text{for } N = 21 \text{ teeth,} \\ 0.106 & \text{for } N = 23 \text{ teeth,} \\ 0.108 & \text{for } N = 25 \text{ teeth,} \\ 0.111 & \text{for } N = 27 \text{ teeth,} \\ 0.114 & \text{for } N = 30 \text{ teeth,} \\ 0.118 & \text{for } N = 34 \text{ teeth,} \\ 0.122 & \text{for } N = 38 \text{ teeth,} \\ 0.126 & \text{for } N = 43 \text{ teeth,} \\ 0.130 & \text{for } N = 50 \text{ teeth,} \\ 0.134 & \text{for } N = 60 \text{ teeth,} \\ 0.138 & \text{for } N = 75 \text{ teeth,} \\ 0.142 & \text{for } N = 100 \text{ teeth,} \\ 0.146 & \text{for } N = 150 \text{ teeth,} \\ 0.150 & \text{for } N = 300 \text{ teeth.} \end{cases} \tag{6.15}$$

For $14\frac{1}{2}°$ full-depth involute teeth, the form factor γ is given by [7]

$$\gamma = \begin{cases} 0.067 & \text{for } N = 12 \text{ teeth}, \\ 0.071 & \text{for } N = 13 \text{ teeth}, \\ 0.075 & \text{for } N = 14 \text{ teeth}, \\ 0.078 & \text{for } N = 15 \text{ teeth}, \\ 0.081 & \text{for } N = 16 \text{ teeth}, \\ 0.084 & \text{for } N = 17 \text{ teeth}, \\ 0.086 & \text{for } N = 18 \text{ teeth}, \\ 0.088 & \text{for } N = 19 \text{ teeth}, \\ 0.090 & \text{for } N = 20 \text{ teeth}, \\ 0.092 & \text{for } N = 21 \text{ teeth}, \\ 0.094 & \text{for } N = 23 \text{ teeth}, \\ 0.097 & \text{for } N = 25 \text{ teeth}, \\ 0.099 & \text{for } N = 27 \text{ teeth}, \\ 0.101 & \text{for } N = 30 \text{ teeth}, \\ 0.104 & \text{for } N = 34 \text{ teeth}, \\ 0.106 & \text{for } N = 38 \text{ teeth}, \\ 0.108 & \text{for } N = 43 \text{ teeth}, \\ 0.110 & \text{for } N = 50 \text{ teeth}, \\ 0.113 & \text{for } N = 60 \text{ teeth}, \\ 0.115 & \text{for } N = 75 \text{ teeth}, \\ 0.117 & \text{for } N = 100 \text{ teeth}, \\ 0.119 & \text{for } N = 150 \text{ teeth}, \\ 0.122 & \text{for } N = 300 \text{ teeth}. \end{cases} \tag{6.16}$$

The transmitted force is a function of $\sigma_0 \gamma$ and is calculated with the Lewis equation

$$F_t = \sigma B p \gamma, \tag{6.17}$$

or

$$F_t = \sigma p^2 k \gamma = \frac{\sigma \pi^2 k \gamma}{P_d^2}. \tag{6.18}$$

A weaker gear will relate to a smaller $\sigma_0 \gamma$ value. For same material the weaker gears will be the smaller ones.

For design problems when the diameters are not known, the stress is calculated with

$$\sigma = \frac{2M_t P_d^3}{k\pi^2 \gamma N},\qquad(6.19)$$

where M_t represents the torque on the weaker gear, the upper limit is $k = 4$, N is the weaker gear number of teeth, which is limited to 15.

The allowable stress is given by

$$\sigma_{al} = \begin{cases} \dfrac{600\,\sigma_0}{V+600} & \text{when } V < 2000 \ \frac{\text{ft}}{\text{min}}, \\[2ex] \dfrac{1200\,\sigma_0}{V+1200} & \text{when } 2000 < V < 4000 \ \frac{\text{ft}}{\text{min}}, \\[2ex] \dfrac{78\,\sigma_0}{\sqrt{V}+78} & \text{when } V > 4000 \ \frac{\text{ft}}{\text{min}}, \end{cases}\qquad(6.20)$$

where V represents the pitch line velocity and σ_0 the strength endurance given by

$$\sigma_0 = \begin{cases} 8000 \text{ psi} & \text{cast iron material,} \\ 12\,000 \text{ psi} & \text{bronze material,} \\ 10\,000\text{--}50\,000 \text{ psi} & \text{carbon steel material.} \end{cases}\qquad(6.21)$$

When the gear diameters are given, the allowable value for the ratio P_d^2/γ is

$$\frac{P_d^2}{\gamma} = \frac{\sigma\,k\pi^2}{F_t},\qquad(6.22)$$

where the transmitted force is $F_t = 2M_t/d$, allowable stress is σ, the upper limit is $k = 4$, and torque on the weaker gear is M_t. The largest diametral pitch gives the most economical design.

References

[1] E.A. Avallone, T. Baumeister, A. Sadegh, Marks' Standard Handbook for Mechanical Engineers, 11th edition, McGraw-Hill Education, New York, 2007.

[2] A. Bedford, W. Fowler, Dynamics, Addison Wesley, Menlo Park, CA, 1999.

[3] A. Bedford, W. Fowler, Statics, Addison Wesley, Menlo Park, CA, 1999.

[4] M. Buculei, D. Bagnaru, Gh. Nanu, D.B. Marghitu, Computing Methods in the Analysis of the Mechanisms with Bars, Scrisul Romanesc, Craiova, 1986.

[5] R. Budynas, K.J. Nisbett, Shigley's Mechanical Engineering Design, 9th edition, McGraw-Hill, New York, 2013.

[6] J.A. Collins, H.R. Busby, G.H. Staab, Mechanical Design of Machine Elements and Machines, 2nd edition, John Wiley & Sons, 2009.

[7] A.S. Hall, A.R. Holowenko, H.G. Laughlin, Schaum's Outline of Machine Design, McGraw-Hill, New York, 2013.

[8] B.G. Hamrock, B. Jacobson, S.R. Schmid, Fundamentals of Machine Elements, McGraw-Hill, New York, 1999.

[9] R.C. Juvinall, K.M. Marshek, Fundamentals of Machine Component Design, 5th edition, John Wiley & Sons, New York, 2010.

[10] K. Lingaiah, Machine Design Databook, 2nd edition, McGraw-Hill Education, New York, 2003.

[11] D.B. Marghitu, Mechanical Engineer's Handbook, Academic Press, San Diego, CA, 2001.

[12] D.B. Marghitu, M.J. Crocker, Analytical Elements of Mechanisms, Cambridge University Press, Cambridge, 2001.

[13] D.B. Marghitu, Kinematic Chains and Machine Component Design, Elsevier, Amsterdam, 2005.

[14] D.B. Marghitu, M. Dupac, H.M. Nels, Statics with MATLAB, Springer, New York, NY, 2013.

[15] D.B. Marghitu, M. Dupac, Advanced Dynamics: Analytical and Numerical Calculations with MATLAB, Springer, New York, NY, 2012.

[16] D.B. Marghitu, Mechanisms and Robots Analysis with MATLAB, Springer, New York, NY, 2009.

[17] C.R. Mischke, Prediction of stochastic endurance strength, Transaction of ASME, Journal Vibration, Acoustics, Stress, and Reliability in Design 109 (1) (1987) 113–122.

[18] R.L. Mott, Machine Elements in Mechanical Design, Prentice-Hall, Upper Saddle River, NJ, 1999.

[19] W.A. Nash, Strength of Materials, Schaum's Outline Series, McGraw-Hill, New York, 1972.

[20] R.L. Norton, Machine Design, Prentice-Hall, Upper Saddle River, NJ, 1996.

[21] R.L. Norton, Design of Machinery, McGraw-Hill, New York, 1999.

[22] W.C. Orthwein, Machine Component Design, West Publishing Company, St. Paul, 1990.

[23] D. Planchard, M. Planchard, SolidWorks 2013 Tutorial with Video Instruction, SDC Publications, 2013.

[24] I.H. Shames, Engineering Mechanics – Statics and Dynamics, Prentice-Hall, Upper Saddle River, NJ, 1997.

[25] J.E. Shigley, C.R. Mischke, Mechanical Engineering Design, McGraw-Hill, New York, 1989.

[26] J.E. Shigley, C.R. Mischke, R.G. Budynas, Mechanical Engineering Design, 7th edition, McGraw-Hill, New York, 2004.

[27] J.E. Shigley, J.J. Uicker, Theory of Machines and Mechanisms, McGraw-Hill, New York, 1995.

[28] A.C. Ugural, Mechanical Design, McGraw-Hill, New York, 2004.

[29] S. Wolfram, Mathematica, Wolfram Media/Cambridge University Press, Cambridge, 1999.

Index

A
Absolute viscosity, 167
Acme thread, 105, 108, 109
Addendum, 193
Addendum circle, 193

B
Ball bearing, 141, 143
   angular contact, 144
   application factor, 148
   contact angle, 144
   radial, 142
   radial-thrust, 142
   thrust, 142
Base circle, 192, 194
Bearing basic number, 144
Bearing reliability factor, 146

C
Castigliano's theorem, 20
Circular pitch, 193
Clamping force, 110, 111
Constant life fatigue diagram, 74
Constant life lines, 74
Curvature, 17

D
Dedendum, 193
Diametral pitch, 194
Dynamic viscosity, 167

E
Endurance
   limit, 69, 79, 80
   strength, 69
Epicyclic, 195

F
Fasteners, 103
Fatigue
   failure, 69
   limit, 70
   strength, 69

G
Goodman criterion, 76
Goodman lines, 74

H
Helix angle, 104
Hydrodynamic lubrication, 170

I
Initial preload, 110
Interference, 194
Involute curve, 191

J
Joint constant, 111

K
Kinematic viscosity, 168

L
Lead angle, 104, 105, 107, 108
Life of a rolling bearing, 146
Load factor, 72

M
Median life, 146
Miner's rule, 75
Module, 194
Mohr's circle, 4, 6
Monograde oil, 169
Multigrade oil, 169

N
Normal stress, 1

P
Petroff's equation, 171
Pitch base, 194
Pitch circle, 192
Pitch diameter, 193, 204
Pitch line velocity, 221
Pitch point, 191, 192, 203

Printed in the United States
By Bookmasters